Generalized Low-Voltage Circuit Techniques for Very High-Speed Time-Interleaved Analog-to-Digital Converters

ANALOG CIRCUITS AND SIGNAL PROCESSING SERIES

Consulting Editor: Mohammed Ismail. *Ohio State University*

For other titles published in this series, go to
http://www.springer.com/series/7381

Sai-Weng Sin · Seng-Pan U · Rui Paulo Martins

Generalized Low-Voltage Circuit Techniques for Very High-Speed Time-Interleaved Analog-to-Digital Converters

 Springer

Sai-Weng Sin
University of Macau
Faculty of Science and Technology
Dept of Electrical and Electronics Engin
Av. Padre Tomas Pereira Taipa
000 Macao
China, People's Republic
terryssw@umac.mo

Seng-Pan U
University of Macau
Faculty of Science and Technology
Dept of Electrical and Electronics Engin
Av. Padre Tomas Pereira Taipa
000 Macao
China, People's Republic
benspu@umac.mo

Prof. Rui Paulo Martins
University of Macau and Tech.
Univ.of Lis
Faculty of Science and Technology
Dept of Electrical and Electronics Engin
Av. Padre Tomas Pereira Taipa
000 Macao
China, People's Republic
rmartins@umac.mo

ISBN 978-90-481-9709-5   e-ISBN 978-90-481-9710-1
DOI 10.1007/978-90-481-9710-1
Springer Dordrecht Heidelberg London New York

Library of Congress Control Number: 2010935169

Cover design: SPi Publisher Services

Printed on acid-free paper

Springer is part of Springer Science+Business Media (www.springer.com)

Preface

Analog-to-Digital Converters (ADCs) play an important role in most modern signal processing and wireless communication systems where extensive signal manipulation is necessary to be performed by complicated digital signal processing (DSP) circuitry. This trend also creates the possibility of fabricating all functional blocks of a system in a single chip (System On Chip – SoC), with great reductions in cost, chip area and power consumption of portable devices which are inevitably large if the corresponding signal-conditioning circuits are implemented with discrete or individual off chip analog elements. However, this tendency places an increasing challenge, in terms of speed, resolution, power consumption, and noise performance, in the design of the front-end ADC which is usually the bottleneck of the whole system, especially under the unavoidable low supply-voltage imposed by technology scaling, as well as the requirement of battery operated portable devices Interleaved Analog-to-Digital Converters.

Interleaved Analog-to-Digital Converters will present new techniques tailored for low-voltage and high-speed Switched-Capacitor (SC) ADC with various design-specific considerations. The book is organized in seven chapters as follows:

Chapter 1 presents an overview of the introductory aspects of the current state-of-the-art low-voltage high-speed ADC designs and it also addresses the motivation and objectives of this research work, besides presenting its originality.

Chapter 2 will discuss the impacts of the CMOS technology scaling in the design of analog circuits, at the device level, with the degradation of both the intrinsic gain and the speed of deep-submicron transistors. Besides that, at circuit level, it presents the design challenges due to the reduction in the analog supply voltage, which leads to the problems of floating switches as well as of reduced voltage headroom for low-voltage opamps. Current solutions to these issues will also be discussed and a comparison between advantages and disadvantages among them will be drawn.

Chapter 3 will present practical considerations of switched-capacitor circuits design within a low-voltage environment, e.g. the common-mode feedback circuits, front-end input interfaces to continuous-time input signal, level-shifting techniques, process-insensitive biasing as well as gain-and-offset compensation. These advanced low-voltage circuit techniques greatly relax the speed-power-accuracy trade-offs which are imposed by the reduced headroom in low-voltage applications.

Chapter 4 will address the concept of Time-Interleaving (TI) which represents an efficient solution for designing very-high-speed circuits and systems under the speed limitations imposed by the technology. On the other hand, since channel mismatches will severely limit the system performance, four different kinds – offset, gain, timing and bandwidth mismatches will be thoroughly analyzed and discussed. Closed-form spectrum expressions, as well as the Signal-to-Noise-and-Distortion Ratio, will be derived for a simplified analysis, at an early design phase, of the performance of time-interleaved systems under various channel mismatches.

Chapter 5 will present the design of a low-voltage 1.2 V, 10 b, 60–360 MS/s reconfigurable time-interleaved analog-to-digital converter which is implemented in 0.18 μm CMOS technology. Design considerations about the various ADC's circuit building blocks, as well as special attention to the layout for high-speed mismatch-insensitive design, will also be addressed.

Chapter 6 will subsequently illustrate the Printed-Circuit Board (PCB) design that supports the experimental testing setup, and will present the measured results of the prototype ADC chip described in Chapter 5. In addition to the measurement summary a comparison with previously reported low-voltage high-speed ADCs will also be provided.

Chapter 7 will finally draw the relevant concluding remarks of this book and proposes prospective future research works.

Appendixes will also be provided to introduce: (i) the operation principle of the proposed common-mode feedback techniques, as presented in Chapter 3; (ii) the mathematic derivation for the SNDR of the bandwidth mismatch in TI-ADCs; (iii) the estimation of noise performance in various building blocks of advanced reset-opamp circuits; and (iv) a mathematic analysis and proof of the special case of gain-mismatch analysis presented in the measurement results of Chapter 6.

<table>
<tr><td>Macau</td><td>Sai-Weng Sin, Terry</td></tr>
<tr><td>May 2010</td><td>Seng-Pan U, Ben</td></tr>
<tr><td></td><td>Rui Paulo Martins</td></tr>
</table>

Acknowledgement

We would like to express our thanks to the Research Committee of University of Macau (UM) as well as to the Macao Science and Technology Development Fund (FDCT) for the financial support to the research work presented in this book. Furthermore, our thanks go also to Dr. Sai-Weng Sin's PhD examination committee members, Prof. Iu Vai Pan (University of Macau), Prof. Han Ying Duo (Tsinghua University, China), Prof. Akira Matsuzawa (Tokyo Institute of Technology, Japan), Prof. Bang-Sup Song (University of California-San Diego, USA) and Prof. Franco Maloberti (University of Pavia, Italy) for their participation in the oral defense and their thorough review of the work. We would like also to thank Prof. Mok Kai Meng, Dr. Vai Mang I, Dr. Mak Peng Un and Dr. Wong Man Chung for their guidance and suggestions, within the graduate study period at UM.

We will always remember good experiences of working with our colleagues in the Analog and Mixed-Signal VLSI Laboratory, of UM, during the same period. Among those, we would like to present our special thanks to Alpha Chio, Meshell Ma, Abner Wei, Julia Zhu, Ivor Chan and Victor Wong for their valuable contributions on various technical points, we've really enjoyed sharing the good atmosphere existing in our research team. In particular, we would like to highlight also here the collaboration of Stephen Ao Ieong and Leo Ng, for providing comprehensive CAD and computing support to us and the whole team while they were not specialized in this area and, simultaneously, were quite busy in their own graduate work.

We owe our gratitude also to our dear friends from Chipidea Microelectronics Macao, now, part of Analog IP group of Synopsys, Charles Chan, Lewis Lei, Dustin Chio, Louis Lao, and Norma Vu for their technical and administrative support. Without their help we won't be able to finish this project timely.

Our families surely deserve the biggest thanks for their love, understanding and continuous support. They were always encouraging us in periods of frustration, while providing great support in home operations when we were absent; this book is dedicated to them.

Contents

Abbreviations

ADC	Analog-to-Digital Converter
W/L	Aspect Ratio
A-DDA	Auxiliary Differential-Difference Amplifier
CRT	Cathode Ray Tube
CQFP	Ceramic Quad Flat-Pack
CLM	Channel Length Modulation
CS	Common Source
CMFB	Common-Mode Feedback
CMOS	Complementary Metal-Oxide Semiconductor
CCPSI	Cross-Coupled Passive Sampling Interface
DUT	Device under Test
DNL	Differential Nonlinearity
DDA	Differential-Difference Amplifier
DEC	Digital Error Correction
DSO	Digital Sampling Oscilloscope
DSP	Digital Signal Processing
DIBL	Drain-Induced Barrier Lowering
ENOB	Effective Number of Bit
ERBW	Effective Resolution Bandwidth
EMC	Electromagnetic-Compatibility
ESL	Equivalent Series Inductance
FBCB	Feedback Current Biasing
FOM	Figure-of-Merit
GBW	Gain-Bandwidth Product
HP	High Performance
IN-OU(IS)	Input Nonuniformly sampled, Output Uniformly played out with Impulses
INL	Integral Nonlinearity
IC	Integrated Circuit
ITRS	International Technology Roadmap for Semiconductors

LCD	Liquid Crystal Display
LO	Local Oscillator
LNA	Low Noise Amplifier
LOP	Low Operating Power
LSTP	Low Standby Power
LV-FGC	Low-Voltage Finite-Gain Compensation
LV-OC	Low-Voltage Offset Compensation
MSPS	Mega Sample per Second
MiM	Metal-insulator-Metal
Mux	Multiplexer
MDAC	Multiplying Digital to Analog Converter
O-CMEC	Output Common-Mode Error Correction
PSD	Power Spectral Density
PCB	Printed-Circuit Board
RO	Reset-Opamp
S/H	Sample and Hold
S/H/R	Sample and Hold and Reset
SNDR	Signal to Noise and Distortion Ratio
SNR	Signal to Noise Ratio
SAR	Successive Approximation Register
SMD	Surface Mount Device
SC	Switched-Capacitor
SC- CMFB	Switched-Capacitor Common-Mode Feedback
SO	Switched-Opamp
SoC	System on Chip
TI	Time-Interleaved
THD	Total Harmonic Distortion
T/R	Track and Reset
UGF	Unity-Gain Frequency
VG-CMFB	Virtual Ground Common-Mode Feedback
VCLS	Voltage Controlled Level Shifting

Chapter 1
Introduction

1.1 Low-Voltage High-Speed Analog-to-Digital Conversion

Driven by continuous downscale of integrated circuit technology, Digital Signal Processing (DSP) and microprocessors constitute the most vital part of the evolution of modern electronic systems. Based on the aggressive diminution of the channel length of transistors the processing speed and the complexity of digital logic circuits have experienced a considerable growth while maintaining low power consumption and allowing the efficient implementation of sophisticated algorithms. The density of transistors in the microprocessor has doubled every 18 months as predicted by Moore's Law [1, 2].

Contrarily to the advancement of digital circuits, the analog circuits can suffer from severe limitations imposed by technology scaling. Following the trend of obtaining highly integrated System-on-Chip (SoC) with low cost, smaller size and superior performance, the analog circuits must be designed with the same technology, on the same substrate, and under the same supply voltage of digital circuits. On the other hand, as technology shrinks the intrinsic transistor gain decreases and the supply voltage drops, as required by deep-submicron operation, leading to increasing design challenges in high-speed high-resolution analog circuits.

Wider complexity and enlarged functionality performed in the digital domain create almost insurmountable pressures to the design of a minimum number of unavoidable analog components. Among which the Analog-to-Digital Converter (ADC) must be considered one of the most important components, since it implements the interface among all the various types of analog signals from our environment (such as the radio, audio and video signals, etc.) and the digital processor, as shown in Fig. 1.1. As a result, the boundary between analog and digital (the place where the ADC should be located) is constantly moving towards the real nature analog world, increasing the complexity of operation in the digital domain and simplifying the tasks to be developed in the analog domain. However, this trend imposes higher stringent requirements in the conversion speed and resolution of the ADC, even before the technology advantages of the DSP could be in full play.

S.-W. Sin et al., *Generalized Low-Voltage Circuit Techniques for Very High-Speed Time-Interleaved Analog-to-Digital Converters*, Analog Circuits and Signal Processing, DOI 10.1007/978-90-481-9710-1_1, © Springer Science+Business Media B.V. 2010

As shown in Fig. 1.2 [3], the required higher resolution at an higher speed ADC always come with a cost of increased power consumption. In fact, ADCs represent often the main bottlenecks of the whole electronic system, determining the operational limits in terms of speed, resolution, noise, area, as well as power consumption, in particular in battery-operated portable devices that impose severe requirements on low operating power and supply voltage (with degraded performance which is clearly demonstrated in Fig. 1.2).

ADCs with reconfigurable capability [4–9] are also desirable to cater for the simultaneously different performance requirements of a wide range of signal bandwidths and applications [4], since multiple ADCs within the same system will lead to larger area, lower level of integration and higher cost, while single ADC with the most stringent application requirements will imply an unnecessary large power consumption. Finally, reconfigurable ADCs also allow a faster time-to-market cycle.

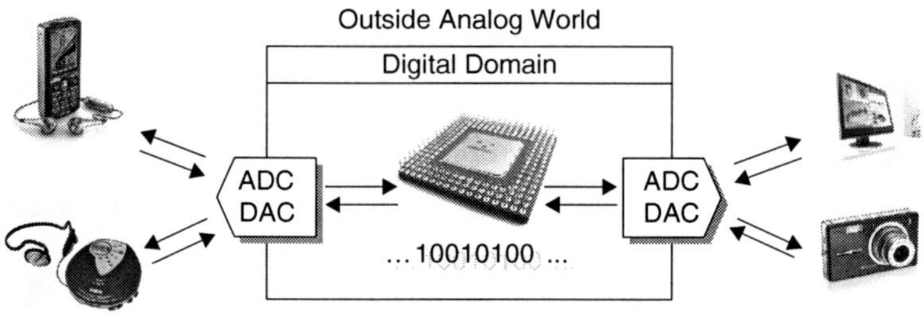

Fig. 1.1 The analog nature of our world

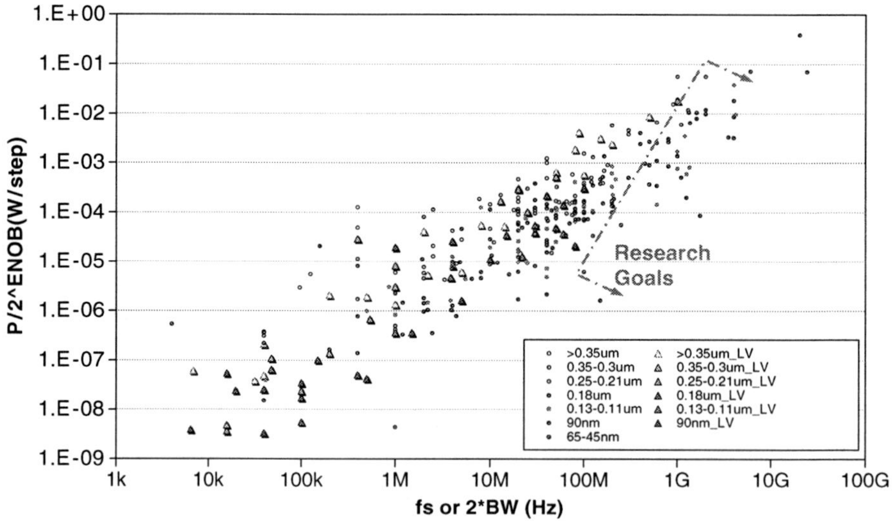

Fig. 1.2 A survey on power per conversion step (Power/2^{ENOB}) versus speed in the state-of-the-art ADCs [3]

1.2 Applications of High-Speed ADCs

High speed ADCs are of paramount importance in a broad area of interests to serve a wide-variety of complex signal processing needs. One good example that confirms this assumption can be represented by the conventional architecture of a super-heterodyne receiver, shown in Fig. 1.3 [10]. The whole process of down-conversion is performed in the analog domain while the ADCs are used to sample the baseband signals. Although in this type of receiver architecture the required conversion rate of the ADCs can be relaxed, it would not be suitable to meet the future trends of single-chip wireless receivers, due to the large amount of off-chip analog processing components, like the image-rejection filter. To overcome this drawback in terms of system integration Fig. 1.4 shows a direct-IF digitizing receiver, in which the RF-to-IF down-conversion is still performed in the analog domain but where the IF-to-Baseband down-conversion is now carried out in the digital domain. Due to the advances experienced by analog-to-digital conversion techniques it is now possible to place the ADC at the IF stage in order to sample directly the signals around the IF frequency [11]. The sampling frequency of the ADC is usually set at four times the IF center frequency to simplify the architecture of the digital local

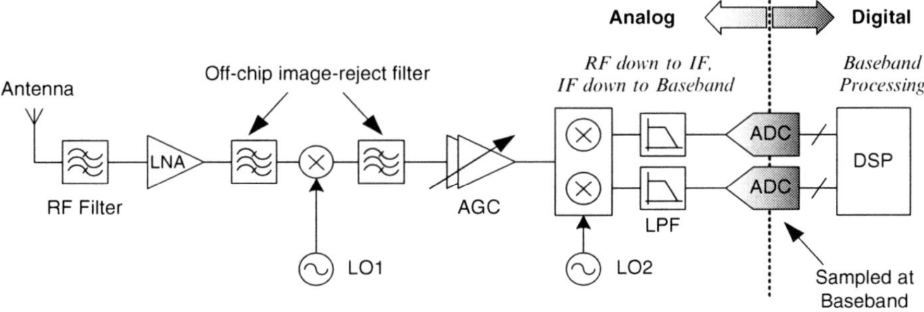

Fig. 1.3 Conventional super-heterodyne receiver architecture

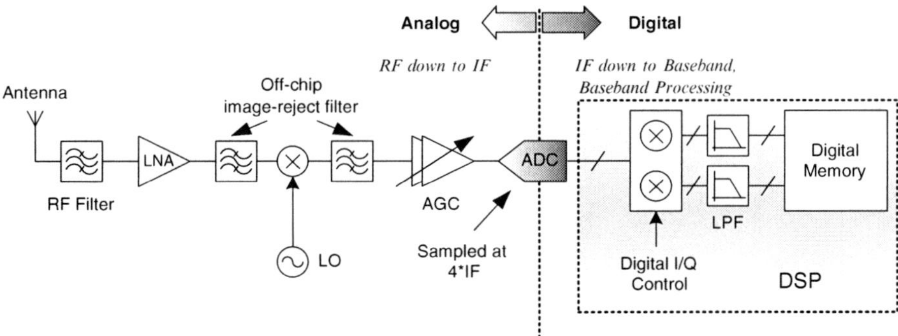

Fig. 1.4 Direct-IF digitizing receiver architecture

oscillator (LO) and the down-conversion process from IF to baseband. Although the required speed of the ADC is quite high (probably in the range of tens to 100 MHz), many of the analog functions are now performed in the digital domain instead leading this receiver architecture to an increased integration capability. By expanding this idea, the ADC can be even placed just after the Low Noise Amplifier (LNA) to sample directly the RF signals, as illustrated in Fig. 1.5, with an architecture known as software radio [12, 13], which is very attractive to cope with the future trend of single-chip wireless receiver mainly because of the elimination of off-chip discrete components. However, it still requires considerable research and development efforts in order to obtain extremely high speed ADCs (probably in the GHz range).

The Digital Sampling Oscilloscope (DSO) is another type of electronic equipment that requires high speed ADCs. It comprises signal conditioning circuitry, a high speed ADC, a buffer memory and a display system (Fig. 1.6) [14, 15]. Many DSOs utilize an

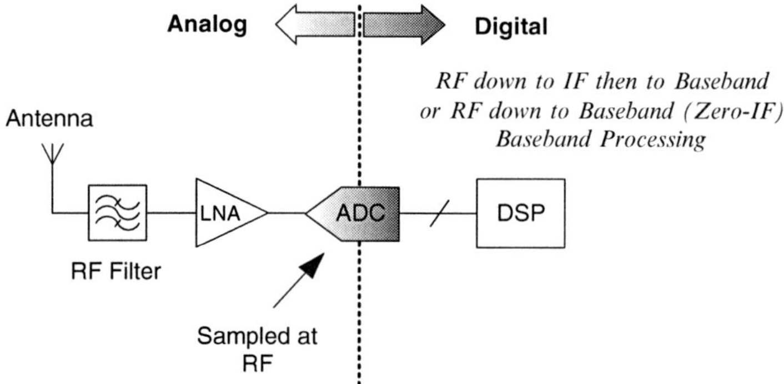

Fig. 1.5 Software radio conceptual architecture

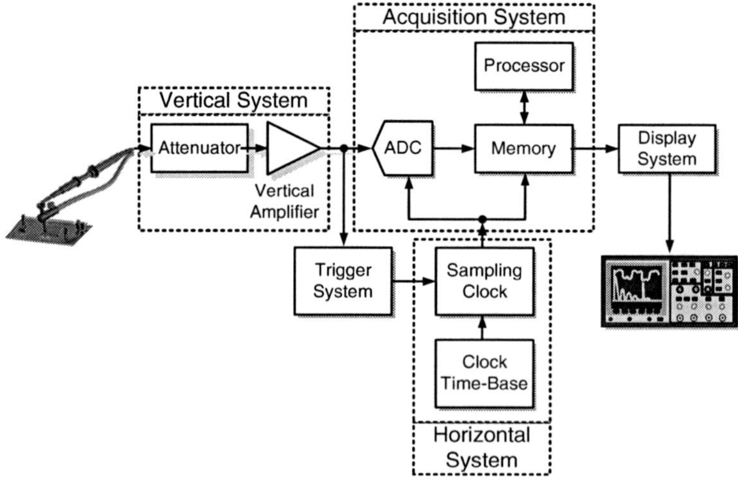

Fig. 1.6 The block diagram of a digital sampling oscilloscope (DSO)

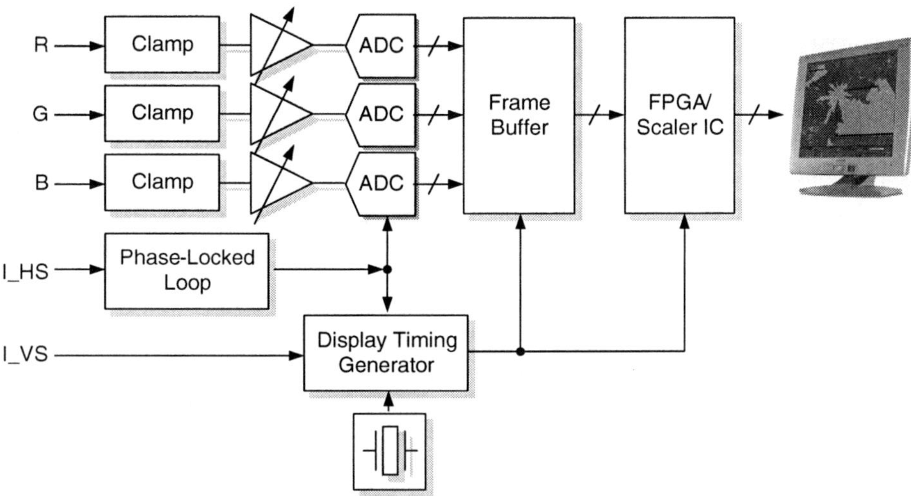

Fig. 1.7 A typical block-diagram of a LCD system front-end

equivalent-time sampling technique, which uses high-speed sampling circuitry with small aperture time to sample very high bandwidth input signals in the GHz range. The sampling clocks of these circuits can be relatively slow, in the order of a few mega-samples per second, turning this technique to be only suitable for narrow band periodic input signals. When broadband or non-periodic signals need to be digitized, very high clock rate ADC using Nyquist rate sampling are necessary, which implies a sample rate greater than twice the bandwidth of the incoming signal. In addition, for high dynamic range measurements a ten to 12 bit resolution of the ADC will be required for the DSO [15].

High speed ADCs are also highly demanded for Liquid-Crystal-Display (LCD) systems. Unlike their Cathode-Ray-Tube (CRT) counterparts LCD monitors need digital driving signal, while many video sources are analog (e.g. the standardized RGB signals). Then, ADCs are required to convert these analog video signals to digital pixel control signals, as shown in Fig. 1.7 [16–18]. Depending on the resolution and refreshing rate of the LCD display, the conversion rate can vary from tens of MSPS (mega-sample per second) to a few hundreds of MSPS with ADCs' resolutions between eight and 12 bits.

1.3 Deep-Submicron CMOS ADCs Designs

The rapid growth of these trends is driving the ADC design towards higher speed, higher resolution, lower supply voltage and power consumption, as well as higher levels of integration with state-of-the-art CMOS technology. As the technology scales down, the speed of the ADC can still be improved due to the reduced intrinsic capacitance of the devices, but it becomes more difficult to maintain the precision of the ADC which relies heavily on the use of switches, high-gain operational amplifiers (opamps) and well matched devices (e.g. capacitors, resistors, and transistors).

Figure 1.8 shows the CMOS technology scaling roadmaps for three different applications of logic technology, as predicted by International Technology Roadmap for Semiconductors (ITRS) [19]. Aggressive technology scaling requires the supply voltages also to be scaled down, but the threshold voltages cannot be proportionally scaled due to limitations associated with leakage currents. As shown in Fig. 1.8, the threshold voltage can be close to 50% of the supply voltage for Low Operating Power (LOP) and Low Standby Power (LSTP) technology, while for High Performance (HP) technology the threshold voltage would be only a small percentage (<20%) of the supply voltage. The HP technology brings additional drawback of a significant amount of leakage currents that increase the standby power and even degrade seriously the performance of the ADCs that rely heavily on the use of switches.

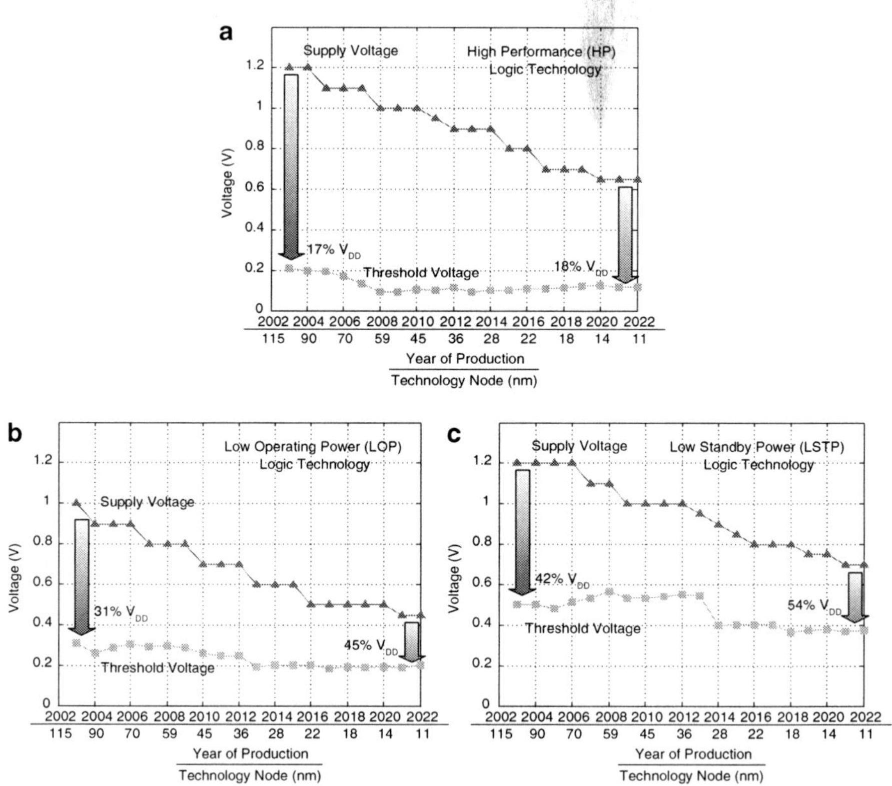

Fig. 1.8 CMOS technology scaling roadmaps for (**a**) High-Performance (HP); (**b**) Low Operating Power (LOP) and (**c**) Low Standby Power (LSTP) logic technology
Source: ITRS'02-07

On the other hand, in deep-submicron CMOS with reduced supply voltage, the output resistances of transistors are generally poorer due to the decreased channel-length and drain-voltage headroom for saturation. The limited supply voltage also imposes difficulties in the design of switches and opamps, thus limiting the usage of common high-precision Switched-Capacitor (SC) circuits. Additionally, the submicron transistors have smaller sizes in order to be faster, which results in poor matching in differential pairs and current mirrors.

Substrate noise [20–23] also becomes a main performance limiter in the design of data converters, which are inherently mixed-signal circuits with sensitive analog and noisy digital blocks sitting on the same substrate. Technology scaling also imposes higher density and complexity of digital circuits, creating noisy spike currents that propagate to analog circuits through the even shorter substrate coupling paths in submicron ICs.

The degraded precision in low-voltage designs are often compensated by using additional circuits that are generally not required in past designs [24, 25] leading to overall performance reduction, e.g. the increase of power consumption, die area, or/and even nibbling up the speed that should be gained by technology scaling. Modern ADCs are needed to meet the same or even higher performance as past designs in sampling linearity, conversion rate, resolution, and power consumption, under a degraded intrinsic low-voltage environment.

1.4 Main Objective and Design Challenges

Based on the previous considerations the main objective of the research that is behind this book is the development of design techniques for a low-voltage high-speed analog-to-digital converter under the most stringent environment imposed by reduced supply voltages and CMOS technology scaling. The ADC will be developed under high threshold voltages that will account for a significant amount of the supply voltages implying that the techniques can be fully scalable and applicable in advanced low-voltage deep-submicron and nano-meter CMOS. The main goals of this work can be summarized as follows:

Investigate and develop a set of generalized circuit techniques for low-voltage SC circuits which behavior is problematic in a low-voltage environment due to the unavailability of floating switches;

Investigate robust solutions for high-speed opamp designs to reduce their sensitivity to process variations, which can help to relax the stringent design margins that are imposed by low-voltage environments;

Explore low-voltage mismatch-insensitive time-interleaved design techniques which are effective in boosting the speed of analog-to-digital converters;

Validate the design techniques with real IC implementations of a low-voltage high-speed ADC design. The main goal would be to extend the operating speed limits of the ADC beyond the trend lines of Fig. 1.2 [3, 6, 24–32] under the continuous reduction of the supply voltage.

The most fundamental and critical challenge of this work lies with the fact that the low-voltage operation turns the floating switch as an unavailable circuit element, thus avoiding the use of many traditional SC design techniques to achieve optimal implementations. Even the simple connection and charging of a capacitor through an opamp becomes problematic, innovative solutions must be developed to overcome them and avoiding as much as possible extra circuitry that degrade the overall power-speed-noise performance trade-offs [24, 25].

Simultaneously low-voltage and high-speed operation also impose intrinsic performance limitations. For example, a reduced overdrive forced by the low supply voltage slows-down the operation of transistors thus reducing the speed of the opamp and increasing the on-resistance of the switches which will seriously degrade the speed of SC circuits. Transistors with larger sizes must be employed in order to maintain the speed with reduced overdrives, which consume extra power, and impose larger parasitic capacitances.

A low-voltage environment with deep-submicron transistors also imposes the reduction in the accuracy of various building blocks. Limited supply voltage prevents the use of cascode transistors such that the use of high-speed single-stage opamps is not possible, implying that two-stage opamps with lower-speed and higher power consumption are generally required to achieve high linearity. Also, current sources without cascoded elements are susceptible to drain-voltage variations, turning the performance of opamps (e.g. gain-bandwidth-product, phase margin and slew-rate) more sensitive to process variations that lead to significant over-design headroom to counteract such problems.

On the other hand, to extend the intrinsic speed limit of the system, time interleaving needs to be adopted as one of the effective approaches to tackle the above-mentioned problems. While ideally time interleaving can enhance linearly the speed associated with the cost of linear increased power and die-area consumption, practical time-interleaved systems possess various types of channel mismatches that create fixed tones or modulation sidebands that limit the overall dynamic range of such systems. Consequently, in time-interleaved ADCs the power-speed-linearity trade-offs would be considerably degraded when compared with single-channel ADCs [33].

Besides the previous sources of difficulties that need to be addressed in the design phase, various other considerations must be carefully undertaken due to the high-speed mixed-signal nature of the integrated systems, as for example the substrate noise coupling, device and path matching, signal and clock routing and shielding, etc. in order to obtain a successful implementation.

References

1. G. Moore, No exponential is forever: but 'Forever' can be delayed!, in *ISSCC Digest of Technical Papers* (Feb 2003), pp. 21–23
2. Intel, Moore's Law. (2003 Online), http://www.intel.com/research/silicon/mooreslaw.htm

3. B. Murmann, ADC performance survey (1997–2007). [Online], http://www.stanford.edu/~murmann/adcsurvey.html
4. K. Gulati, Hae-Seung Lee, A Low-Power Reconfigurable Analog-to-Digital Conver-ter, in *IEEE J. Solid-State Circuits*. **36**(12), 1900–1911 (Dec 2001)
5. P. Nuzzo, G. Van der Plas, F. De Bernardinis, L. Van der Perre, A 10.6mW/0.8pJ power-scalable 1 GS/s 4b ADC in 0.18μm CMOS with 5.8GHz ERBW, in *Proceedings of Design Automation Conference* (July 2006), pp. 873–878
6. I. Ahmed, D. Johns, A 50MS/s (35mW) to 1kS/s (15μW) power scalable 10b pipelined ADC with minimal bias current variation, in *ISSCC, Digest of Technical Paper*, pp. 280–281, 598, (Feb 2005)
7. J. Craninckx, G. Van der Plas, A 65fJ/conversion-step 0-to-50MS/s 0-to-0.7mW 9b charge-sharing SAR ADC in 90nm digital CMOS, in *ISSCC, Digest of Technical Paper* (Feb 2007), pp. 246–247, 600
8. P. Nuzzo, G. Van der Plas, F. De Bernardinis, L. Van der Perre, A re-configurable 0.5V to 1.2V, 10MS/s to 100MS/s, low-power 10b 0.13μm CMOS pipeline ADC, in *Proceedings of Custom Integrated Circuit Conference (CICC)* (Sept 2004), pp. 259–262
9. B. Xia, A. Valdes-Garcia, E. Sanchez-Sinencio, A configurable time-interleaved pipeline ADC for multi-standard wireless receivers, in *Proceedings of European Solid-State Circuit Conference* (Sept 2007), pp. 185–188
10. T. Lee, *The Design of CMOS Radio Frequency Integrated Circuits* (Cambridge University Press, Cambridge, 1998)
11. M. Waltari, L. Sumanen, T. Korhonen, K. Halonen, A self-calibrated pipeline ADC with 200MHz IF-sampling frontend, in *ISSCC, Digest of Technical Paper* (Feb 2002), pp. 314, 469
12. D. Leenaerts, J. van der Tang, C. Vaucher, *Circuit Design for RF Transceivers* (Kluwer, Boston, MA, 2001)
13. L. Weidong, Y. yan, Software radio: technology and implementation, in *Proceedings of 1998 IEEE International Conference on Communication Technology*, vol. 1 (Oct 1998), pp. 1–5
14. K. Rush, P. Byrne, A 4GHz 8b data acquisition system, in *IEEE International Solid State Circuits Conference* (Feb 1991), pp. 176–177
15. W.T. Colleran, *A 10-b, 100MS/s A/D converter using folding, interpolation, and analog encod-ing*, Ph.D. Dissertation, University of California Los Angeles, Los Angeles, CA, Dec 1993
16. H. Marie, P. Belin, R, G, B acquisition interface with line-locked clock generator for flat panel display. in *IEEE J. Solid-St. Circ.* **33**(7), 1009–1013 (July 1998)
17. Texas Instrument, Techniques for sampling high-speed graphics with lower-speed A/D con-verters. Analog Appl. J. (Nov 1999)
18. S.-C. Lee, Y.-D. Jeon, J.-K. Kwon, J. Kim, A 10-bit 205-MS/s 1.0-mm2 90-nm CMOS pipeline ADC for flat panel display applications. in *IEEE J. Solid-St. Circ.* **42**(12), 2688–2695 (Dec 2007)
19. International Technology Roadmaps for Semiconductors (ITRS), *ITRS 2002–2007 Edition Reports - Process Integration, Devices & Structures*. [Online], http://www.itrs.net/reports.html
20. T.J. Schmerbeck, Noise coupling in mixed signal ASICs, in *Low-Power HF Microelectronics: A Unified Approach (Chapter 10)*, ed. by G. Machado (IEE Press, New York, 1996)
21. AMS Analog Group, *Crosstalk in Mixed-Signal Systems* (AMS, 1996)
22. T. Blalack, *Switching Noise in Mixed-Signal Integrated Circuits*, Ph.D. Dissertation, Stanford University, Stanford, CA, 1997
23. B.R. Stanisic, N.K. Verghese, R.A. Rutenbar, L.R. Carley, D.J. Allstot, Addressing substrate coupling in mixed-mode IC's: simulation and power distribution synthesis. in *IEEE J. of Solid-St. Circ.* **29**(3), 226–237 (March 1994)
24. J. Li, G. Ahn, D. Chang, U. Moon, A 0.9-V 12-mW 5-MSPS algorithmic ADC with 77-dB SFDR. in *IEEE J. Solid-St. Circ.* **40**(4), 960–969 (April 2005)
25. P.Y. Wu, V.S.L. Cheung, H.C. Luong, A 1-V 100-MS/s 8-bit CMOS switched-opamp pipelined ADC using loading-free architecture. in *IEEE J. Solid-St. Circ.* **42**(4), 730–738 (April 2007)

26. M. Boulemnakher, E. Andre, J. Roux, F. Paillardet, A 1.2V 4.5mW 10b 100MS/s Pipeline ADC in a 65nm CMOS, in *Digest of Technical Papers of IEEE International Solid-State Circuits Conference (ISSCC)* (Feb 2008), pp. 250–251, 611

27. M. Yoshioka, M. Kudo, T. Mori, S. Tsukamoto, A 0.8V 10b 80MS/s 6.5mW Pipelined ADC with regulated overdrive voltage biasing, in *Digest of Technical Papers of IEEE International Solid-State Circuits Conference (ISSCC)* (Feb 2007), pp. 452–453, 614

28. D.J. Huber, R.J. Chandler, A.A. Abidi, A 10b 160MS/s 84mW 1V subranging ADC in 90nm CMOS, in *Digest of Technical Papers of IEEE International Solid-State Circuits Conference (ISSCC)* (Feb 2007), pp. 454–455, 615

29. B. Hernes, J. Bjørnsen, T.N. Andersen, A. Vinje, H. Korsvoll, F. Telstø, A. Briskemyr, C. Holdø, Ø. Moldsvor, A 92.5mW 205MS/s 10b pipeline IF ADC implemented in 1.2V/ 3.3V 0.13µm CMOS, in *Digest of Technical Papers of IEEE International Solid-State Circuits Conference (ISSCC)* (Feb 2007), pp. 462–463, 615

30. Y. Shimizu, S. Murayama, K. Kudoh, H. Yatsuda, A split-load interpolation-amplifier-array 300MS/s 8b subranging ADC in 90nm CMOS, in *Digest of Technical Papers of IEEE International Solid-State Circuits Conference (ISSCC)* (Feb 2008), pp. 552–553, 635

31. S. Junhua, P. Kinget, A 0.5V 8bit 10Msps pipelined ADC in 90nm CMOS, in *Digest of Technical Papers of IEEE Symposium on VLSI Circuits Conference (VLSI)* (June 2007), pp. 202–203

32. K.-J. Lee, E.-S. Shin, H.-S. Yang, J.-H. Kim, P.-U. Ko, I.-R. Kim, S.-H. Lee, K.-H. Moon, J.-W. Kim, A 90nm CMOS 0.28mm2 1V 12b 40MS/s ADC with 0.39pJ/Conversion-Step, in *Digest of Technical Papers of IEEE Symposium on VLSI Circuits Conference (VLSI)* (June 2007), pp. 198–199

33. C.-K.K. Yang et al., A serial-link transceiver based on 8-G samples/s A/D and D/A converters in 0.25-µm CMOS. in *IEEE J. Solid-St. Circ.* **36**(11), 1684–1692 (Nov 2001)

Chapter 2
Challenges in Low-Voltage Circuit Designs

2.1 Introduction

Low voltage designs continue to play important roles, as well as creating challenges, in modern analog and mixed-signal integrated circuits (IC) that are expected to operate under low supply voltage. This is either due to the technology scaling into sub-100 nm CMOS, or in the context of low-power battery-operated devices that force the ICs to function less than the nominal supply voltage at the specific technology node [1–13]. Both situations create stringent requirements in analog and mixed-signal IC designs due to low-supply voltage headroom, especially when the supply voltage is lower than the nominal value of that technology, e.g. an analog IC designed in 0.18 μm CMOS must tolerate the reduced voltage headroom of a 1.2 V supply voltage while it cannot benefit from smaller parasitics and higher speed of more advanced CMOS technologies, like 130 or 90 nm.

As the feature sizes of CMOS continue to scale down, the performance of analog circuits are degraded due to (i) aggravation of the intrinsic performance of devices as a result of various types of short-channel effects and (ii) design headroom reduction induced by low supply voltage, affecting the normal operation of fundamental building blocks like opamps and switches. All of these aspects will be briefly discussed in this chapter.

2.2 The Impact of CMOS Technology Scaling

The feature sizes of CMOS technology are continuously scaling down as predicted by Moore's Law [14] implying that the supply voltage must also be lowered to incorporate a thinner layer of gate oxide. Digital circuits become more powerful in terms of area and power consumption and can fully benefit from scaling in both feature sizes and supply voltages, since the power consumption of digital circuits is

S.-W. Sin et al., *Generalized Low-Voltage Circuit Techniques for Very High-Speed Time-Interleaved Analog-to-Digital Converters*, Analog Circuits and Signal Processing, DOI 10.1007/978-90-481-9710-1_2, © Springer Science+Business Media B.V. 2010

inversely proportional to the supply voltage V_{DD} and the loading capacitance C_L (also smaller due to the reduced feature sizes):

$$P_{digital} \approx f_{clk}C_L V_{DD}^2 \tag{2.1}$$

where f_{clk} represents the clock frequency. Thus, digital circuits can achieve simultaneously higher speed and lower power consumption in deep sub-micron CMOS technology.

Analog designs cannot completely benefit from technology scaling, in particular, due to the reduction in signal dynamic range imposed by the limited supply voltage, as well as the fact that the threshold voltage cannot be scaled down as fast as the supply voltages, as shown in Fig. 1.8. This creates a challenge when designing an operational amplifier (opamp) basically because of the inadequate design headroom in terms of overdrive ($V_{ov} = V_{GS} - V_{th}$) and the amount of saturation voltage ($V_{DS} - V_{ov}$), while the signal swing must be maximized to maintain the Signal-to-Noise Ratio (SNR). For mixed-signal circuits and systems that are implemented with SC circuits, the situation is even problematic when the switches cannot be turned on as the supply voltage is lowered.

The effects of reduced SNR in the performance of analog circuits can be analyzed quantitatively. Assuming that the opamp is designed with a two-stage architecture that has the best achievable output swing with only two stacked Common-Source (CS) output stage, the opamp output swing or the peak-to-peak signal amplitude can be derived as:

$$V_{signal,pp} = V_{DD} - 2V_{ov} \tag{2.2}$$

The power consumption of the opamp is approximately proportional to the tail current I_{SS} of the differential-pair:

$$P_{opamp} \propto V_{DD}I_{SS} \tag{2.3}$$

The thermal noise power is inversely proportional to the capacitive load. If the speed of the opamp is maintained, then the transconductance g_m is also proportional to the capacitive load. If the proportional increase of the aspect ratio (W/L) of the differential-pair is also allowed in order to maintain a constant overdrive voltage (which is usually true especially in low-voltage designs), then the thermal noise power can be derived as:

$$P_{noise} = kT/C \propto 1/g_m = V_{ov}/(2I_D) \propto 1/I_{SS} \tag{2.4}$$

where k is the Boltzmann constant, T is the temperature, and $I_D = I_{SS}/2$ is the drain current of the differential pair. Finally, the SNR can be evaluated as

$$SNR = \frac{P_{signal}}{P_{noise}} \propto \frac{(V_{DD} - 2V_{ov})^2/8}{1/I_{SS}} \approx P_{opamp}V_{DD}/8 \tag{2.5}$$

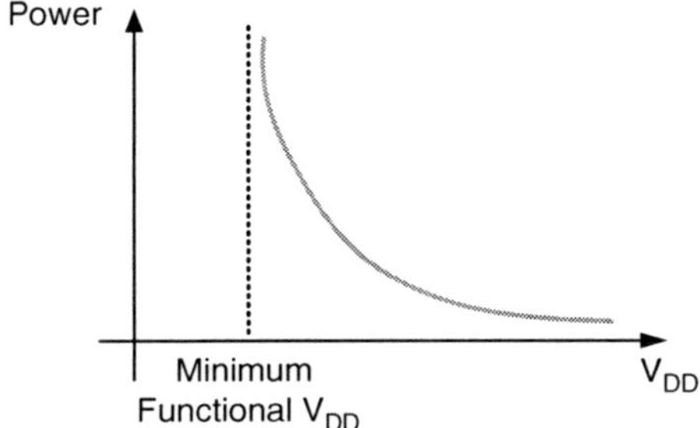

Fig. 2.1 Power consumption versus supply voltage in practical analog circuits

or

$$P_{opamp} \propto \frac{SNR}{V_{DD}} \tag{2.6}$$

showing that to maintain the performance (SNR) the power consumption of analog circuits will be increased due to the reduction in supply voltage from technology scaling. In practice due to the requirements of opamp and switch designs there will always exist a minimum V_{DD} to ensure the analog circuit is fundamentally functional, and the analog circuit will fail to work even with very large power when VDD is below this minimum value, as shown in Fig. 2.1.

2.3 Design Challenges: Intrinsic Performance Degradation

As the technology is scaled down and the minimum channel length of transistors continuous to shrink one of the apparent performance benefits is the increase in the speed of devices due to the decrease in device area. However, using deep sub-micron CMOS transistors in the design of analog circuits is not a trivial task due to the fact that the overall performance of the devices is degraded by various types of short-channel effects.

2.3.1 Transconductance Degradation

As gate oxide thickness and also the channel length scale down the electric field goes stronger which would lead to mobility degradation and velocity saturation effects [15, 16]. With the consideration of both effects the drain current of the MOSFET can be expressed as

$$I_D = \frac{1}{2} \frac{\mu_0 C_{ox}}{1 + \left(\frac{\mu_0}{2v_{sat}L} + \theta\right)(V_{GS} - V_{TH})} \frac{W}{L} (V_{GS} - V_{TH})^2$$

$$\approx \begin{cases} \frac{1}{2}\mu_0 C_{ox}\frac{W}{L}(V_{GS}-V_{TH})^2, \text{ for } \left(\frac{\mu_0}{2v_{sat}L}+\theta\right)(V_{GS}-V_{TH})<<1 \\ \frac{1}{2}\dfrac{\mu_0 C_{ox}}{\left(\dfrac{\mu_0}{2v_{sat}L}+\theta\right)}\frac{W}{L}(V_{GS}-V_{TH}), \text{ for } \left(\frac{\mu_0}{2v_{sat}L}+\theta\right)(V_{GS}-V_{TH})>>1 \end{cases} \quad (2.7)$$

where μ_0 is the mobility, Cox is the gate-oxide unit capacitance, θ is the fitting parameter accounting for the mobility degradation effect, and v_{sat} is the saturation velocity. When L becomes small and θ becomes large (due to thinner gate-oxide) the drain current tends to deviate from the traditional square-law to linear characteristic. This can be clearly illustrated by Fig. 2.2a and the consequence is that transconductance of transistors ceases to augment with increasing overdrive voltage as shown in Fig. 2.2b, degrading the gain and the speed of short-channel transistors.

2.3.2 Output Resistance Degradation

Another obvious short-channel effect is the output resistance degradation and subsequently the intrinsic gain reduction of the transistor. This is not only due to the decreasing channel length but also as a result of other second-order effects. At low value of V_{DS}, the output resistance r_{ds} grows with increasing V_{DS} as governed by the Channel-Length Modulation (CLM) effect that is shown in Fig. 2.3 [17].

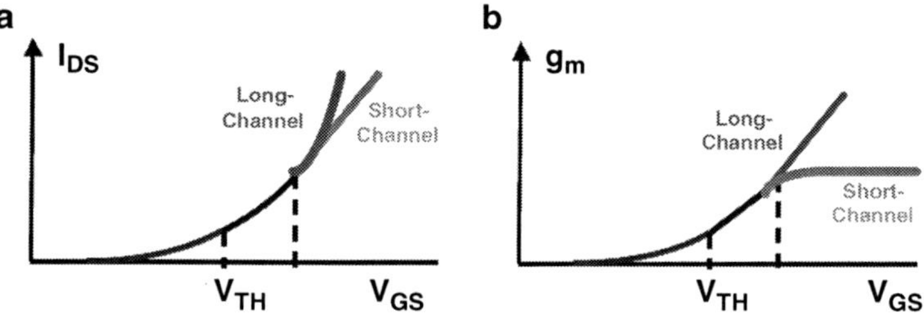

Fig. 2.2 (a) I_D and (b) g_m reduction in deep sub-micron CMOS technology

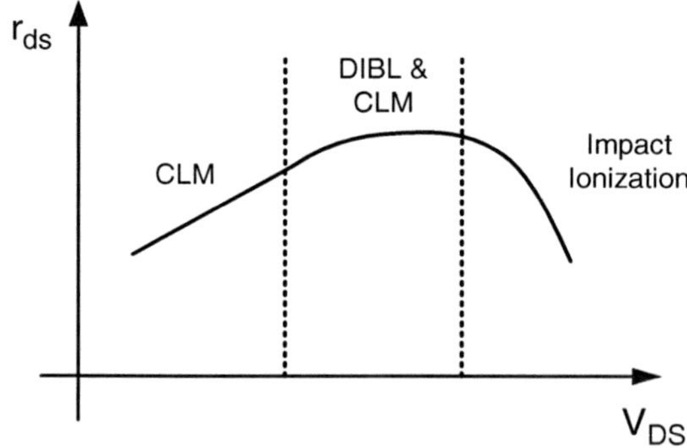

Fig. 2.3 Overall output resistance variation as a function of V_{DS}

However in short-channel devices the continuously increasing V_{DS} will lead to a Drain-Induced Barrier Lowering (DIBL) effect that reduces the threshold voltage thus increasing the drain current. This effect roughly cancels out CLM and also leads to a relatively constant output resistance. As V_{DS} reaches a high value, impact ionization near the drain produces large current flowing into substrate thus further reduces the output resistance.

2.3.3 Trends of Unity-Gain Frequency in Technology Scaling

The main benefit of technology scaling is the increase in the device's operating speed. Unity-Gain Frequency f_T can be used to describe the intrinsic speed of the transistors and can be defined as

$$f_T = \frac{g_m}{2\pi C_{GS}} \approx \frac{\mu C_{ox}(W/L)(V_{GS} - V_{th})}{2\pi \frac{2}{3} WLC_{ox}} = \frac{3\mu(V_{GS} - V_{th})}{4\pi L^2} \tag{2.8}$$

where C_{GS} is the gate-source capacitance. As the feature size scales down the intrinsic speed of the devices increases quadratically. Nevertheless in a short-channel transistor the trend has been slowed down by the velocity saturation effect:

$$f_T \approx \frac{3v_{sat}}{2\pi L} \tag{2.9}$$

This means that the speed increases only linearly with decreasing feature size as shown in Fig. 2.4 [17].

2.4 Circuit Level Design Challenges: Opamps

The reduction in the supply voltage limits the headroom of the operational amplifiers. Cascode structures are not suitable in low-voltage designs and, in order to

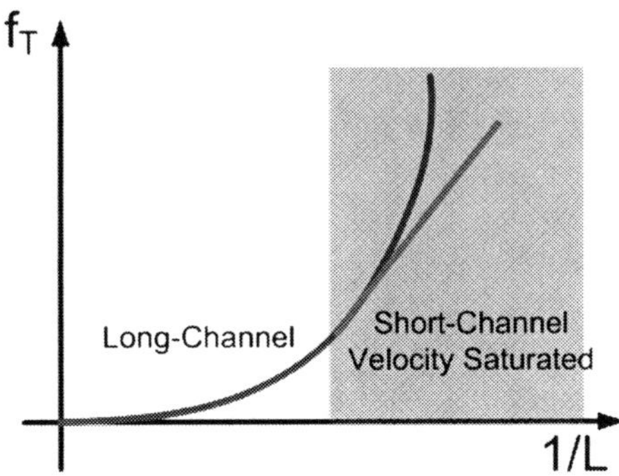

Fig. 2.4 A plot of f_T versus $1/L$ for long- and short-channel devices

achieve high-gain, two-stage opamps are usually used with three stacked transistors in the input stages and two in the output stages, as shown in Fig. 2.5, leading to lower speed and larger power consumption.

On the other hand, Fig. 2.6 exhibits the architecture of a low-voltage two-stage opamp [2, 3] that is widely-used in modern low-voltage designs. The input and output stages from Fig. 2.5 are used to maximize the input and output swing. The input stage utilizes an approach similar to a folded-cascode design to fold the CM level of the differential pair output to a more appropriate level (by M4A,B) suitable for biasing the output common-source stage. However the gain of the first stage is still in the order of $g_m r_o$, since the upper transistors M5A,B is not cascoded due to the supply headroom limitation. The overall gain of the two-stage opamp is in the order of $(g_m r_o)^2$.

Designing Common-Mode Feedback (CMFB) for a two-stage opamp is always more difficult than a single-stage opamp since it would be necessary to stabilize the CM level of two pairs of high-impedance nodes, namely the drains of M4A, M4B, M6A and M6B. Simultaneously controlling the output CM level of two stages requires a feedback point located in the first stage, but V_{CMFB} cannot be directly applied to the gate of any first stage's current source transistors due to the wrong polarity, which will lead to positive feedback in the CM loop. To solve this

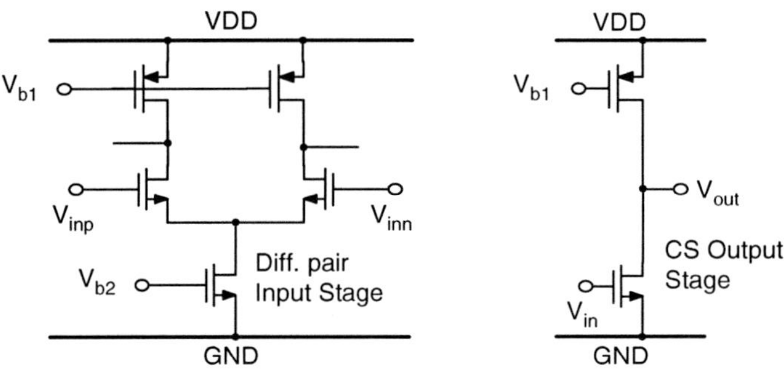

Fig. 2.5 The input and output stages of a low-voltage two-stage opamp

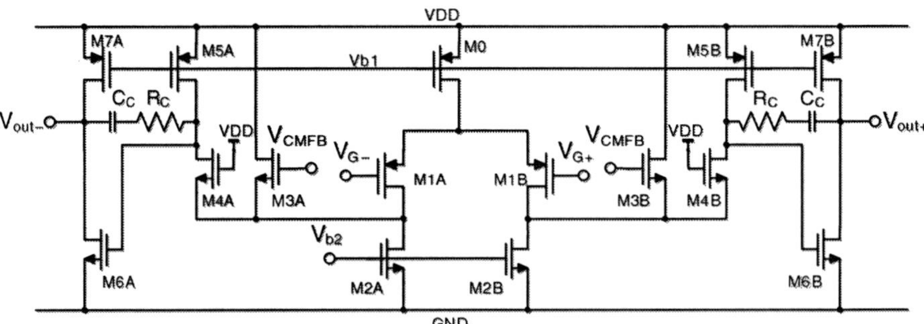

Fig. 2.6 A widely-used low-voltage two-stage opamp

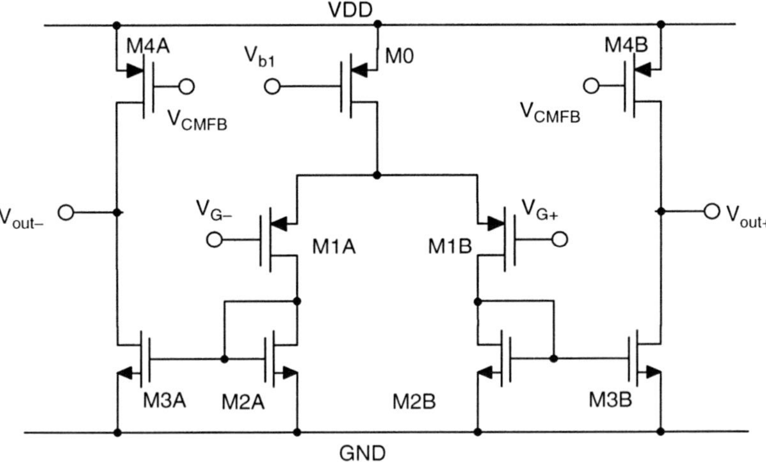

Fig. 2.7 A low-voltage single-stage current-mirror opamp

problem, an additional pair of NMOS current sources (M3A, M3B) with the source tied to the folding node of the first stage needs to be added to provide CMFB control on both stages with proper polarity [2].

Due to the low voltage headroom, this two-stage opamp cannot achieve high speed since the internal high-impedance node needs miller compensation, and the power is large as a result of extra current branches in the folding path and the CMFB controlling branch. On the other hand, a single-stage current-mirror opamp, as shown in Fig. 2.7, also uses the input and output stages of Fig. 2.5, which can achieve potentially higher speed than the two-stage opamps. However, without the use of cascoded devices the opamp from Fig. 2.7 can only achieve a very low value of DC gain (in the order of $g_m r_o$), which is not adequate in modern high-resolution data converter designs. It is clear from this point of view that designing high-speed, high-resolution analog circuits in low-voltage environment is not a trivial task.

2.5 Circuit Level Design Challenges: Switches

Designing SC circuits in a low-voltage environment is especially problematic when comparing it with other analog circuits due to the large dependency on the operation of the switches. When the supply voltage is less than the sum of the threshold voltages of NMOS and PMOS ($V_{DD} \leq V_{thn} + |V_{thp}|$), floating switches that are designed to pass signals with large swings cannot be turned on as shown in Fig. 2.8 [13]. Even if they can be switched on, a large distortion will appear as a result of signal-dependent on-resistance imposed by very small gate overdrive voltage as expressed in (2.10), for NMOS transistors:

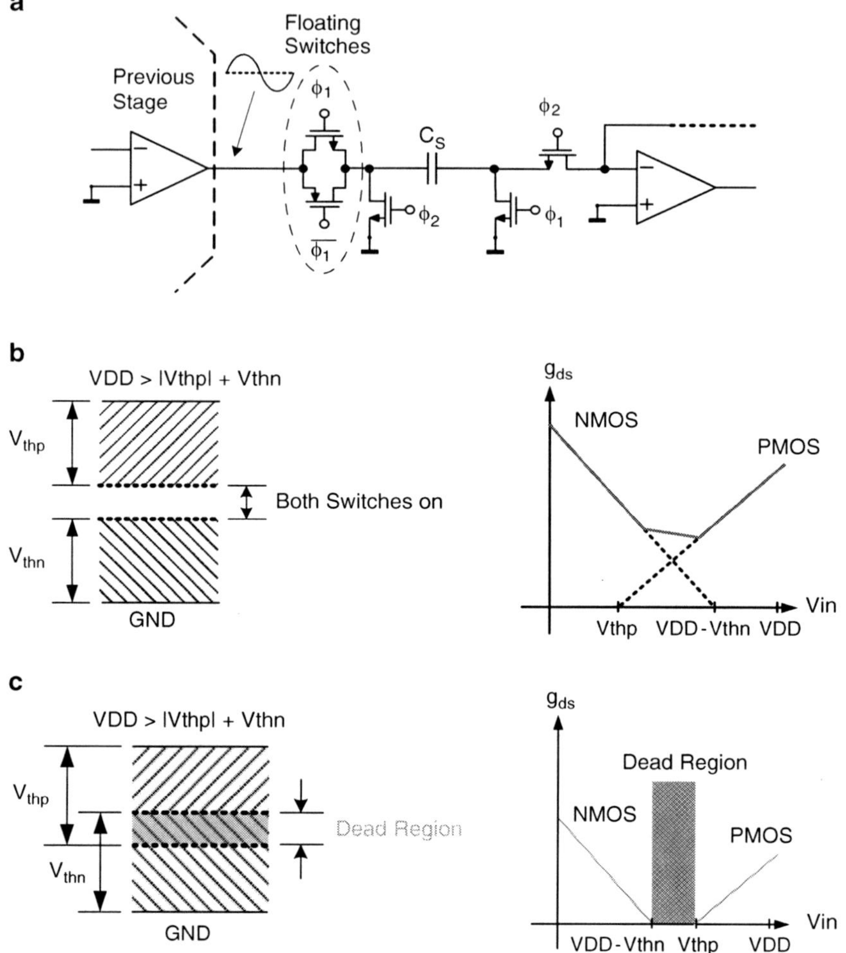

Fig. 2.8 (**a**) A floating switch passing large signal swing and the conductance of the switch in (**b**) a nominal voltage design and in (**c**) a low-voltage design

$$R_{on} = \frac{1}{\mu_n C_{ox}\left(\frac{W}{L}\right)(V_{DD} - V_{in} - V_{th})} \tag{2.10}$$

where R_{on}, μ_n, C_{ox}, W/L, V_{DD}, V_{in} and V_{th} represent the on-resistance, mobility, oxide unit capacitance, transistor aspect ratio, supply voltage, input voltage and threshold voltage, respectively. More accurately, floating switches failed to have reasonable on-resistances, when the following criterion is satisfied:

$$V_{DD} < V_{thn} + \left|V_{thp}\right| + 2V_{ov} \tag{2.11}$$

which should be used as a main indicator of a truly low-voltage design.

In modern low-voltage designs various techniques are available to overcome the problem of turning-on the floating switches as presented next.

2.5.1 Switch Positioning

One of the first principles to be adopted in low-voltage design is to avoid the use of floating switches whenever possible by carefully positioning their location. Switches that pass small signal swings, such as those connected to constant DC voltage (e.g. supply rails and CM voltages) as well as virtual grounds as shown in Fig. 2.9, are not considered as floating switches, and they can always be turned on by biasing those DC voltages and virtual ground CM voltages to near supply rails (the resulting problem of CM level difference can be solved by level-shifting techniques as discussed next in Chapter 3). The most problematic switches are those that pass large signal swing, e.g. the foremost front-end sampling switches and the switches at the opamp output nodes (Fig. 2.8a) that should be minimized whenever possible.

2.5.2 Clock Boosting

The number of floating switches can be often minimized but they still cannot be completely avoided by the operating principles of SC circuits. The reason why the floating switches cannot be turned on is due to the insufficient overdrive voltage in low-voltage environment. One of the solutions is to use a high voltage clock (e.g., $2V_{DD}$ clock) to drive the floating switches, as illustrated in Fig. 2.10a [7, 18]. With this technique the transmission gate can always be switched on (as shown in the Fig. 2.10b) and traditional SC circuit techniques can be utilized since all floating switches problems mentioned before will not arise. However, due to the on-chip high voltage level, there will be a long-term reliability concern since the oxide

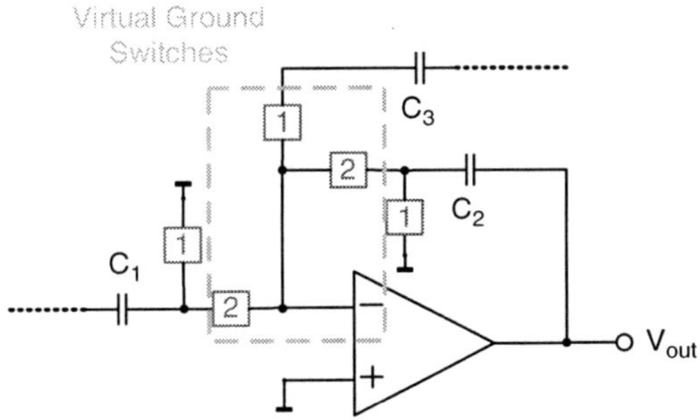

Fig. 2.9 Switches that have small signal swings

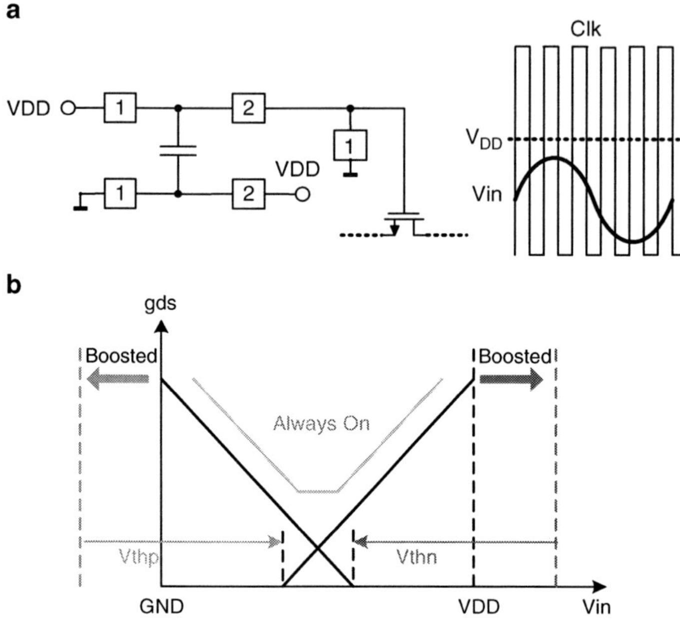

Fig. 2.10 The principles of clock boosting: (**a**) the functional diagram and (**b**) the combined conductance of the floating transmission gate

cannot sustain the voltage stress that exceeds the maximum V_{DD} specified by a technology process. Also, nonlinearity still exists since the on-resistance depends on the signal voltage.

2.5.3 Bootstrapped Switches

The long-term reliability issue due to higher voltage stress over gate oxide has motivated the research for a more effective approach to drive the floating switches. Bootstrapped circuits as shown in Fig. 2.11 can be used to solve the problem [8–11, 19–22]. The operation principle of the circuit is to create a constant gate-source voltage (equal V_{DD}) across the floating-switch transistor such that it has enough overdrive regardless of the input voltage. Also the nonlinearity is suppressed since this technique creates a constant on-resistance independent of the input signal level. As a result, it does not require a transmission gate and only one type of switch would be needed. Since the floating switch is bootstrapped the traditional SC circuit structures can be used in a similar way as the clock boosting technique. However, each floating switch will have a separate bootstrapped circuit which consumes additional power.

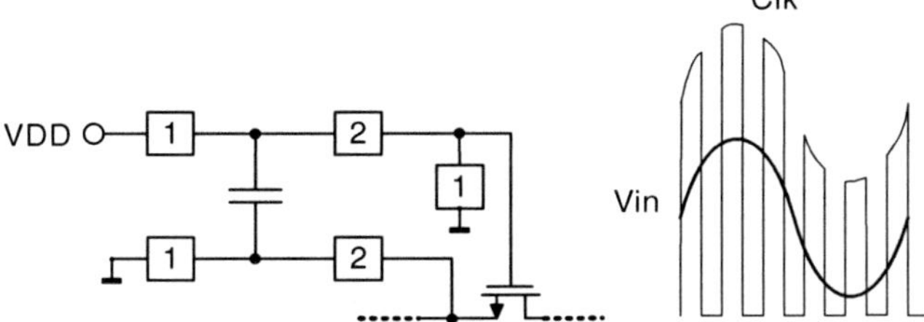

Fig. 2.11 The principle of bootstrapped switch

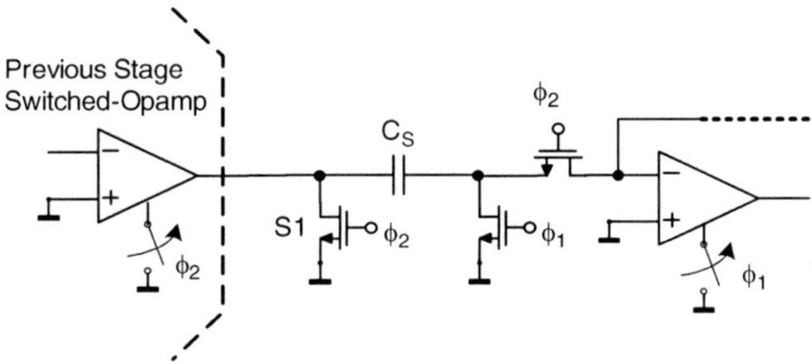

Fig. 2.12 The operation principle of the switched-opamp circuit

2.5.4 Switched-Opamp

The floating switches located at the output of the opamps can also be avoided by using the opamps to simulate switching function. Switched-Opamp (SO) is one of such techniques and its principle is shown in Fig. 2.12 [1, 2, 11, 23–28]. In phase 1, the output voltage of the opamp from previous stage is sampled by the sampling capacitor C_S. In phase 2, the previous stage's opamp is switched-off (by cutting off its biasing current), creating an high-impedance state at the opamp output such that this node can be pulled-down to ground by switch S1 to discharge the sampling capacitor.

Figure 2.13 shows a simple implementation of a switched opamp. The biasing current of the whole opamp is cut-off by the switches S2 and S3. Notice that switched-opamps have an inherited speed limitation since when the opamp is switched-off the output is pulled to ground (by S1 in Fig. 2.12) and the compensation capacitor is discharged. In addition, switch S3 also discharges the gate capacitance of the biasing node of the current sources. This discharging and recharging cycle of SO circuits make them not suitable for high-speed applications. Although later implementations [11, 26–28] have solved most of these disadvantages and the speed has been improved, however SO circuits still suffer from speed degradation due to unavoidable switching-on transients.

The implementation of the SO circuit does not require any on-chip high voltage thus it does not exhibit long-term reliability issues and it can be treated as a truly

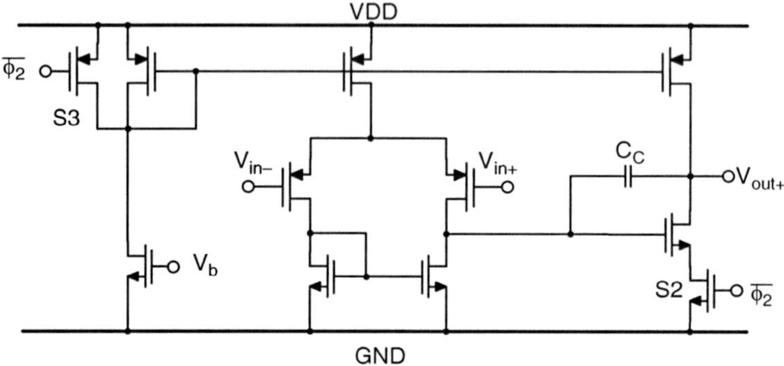

Fig. 2.13 A simple implementation of a switched opamp

low-voltage technique that is compatible with future CMOS technology scaling. No
floating switches are necessary and the linearity is also improved since the on-
resistances of all switches are now signal-independent. Furthermore, the power
consumption of the SO circuit is inherently low since the opamp is switched off half
of the time. SO circuits can be implemented in fully-differential architectures since
in phase 1 the opamp's CM is pulled to GND and thus the output voltage is well
defined, besides that switched-capacitor CMFB can also be applied in SO [2].
However, the limitation of the SO circuit is the slow turn-on transient; In additions,
since no floating switches can be used and with the sampling capacitor C_S always
connected to opamp's output node this limits the utilization of many traditional SC
circuit structures which are well optimized, leading to further degradation in
performance. Finally, SO circuits require extra effort in designing front-end
sampling circuits that interface with the continuous-time input signal since there
will not be a previous stage's opamp which could be switched off at the foremost
front-end stage.

2.5.5 Reset-Opamp

Reset-opamp (RO) (Fig. 2.14) is another low-voltage technique that uses opamps to
simulate floating switches [3, 5, 12, 13]. It relies on unity-gain reset opamp in the
reset-phase to discharge the sampling capacitor which can achieve faster speed
since the opamps are always active. Similar to the SO technique the ROs do not
require generation of on-chip high values of voltage and they are truly compatible
with future low-voltage deep-submicron CMOS processes.

The RO technique exhibits similar advantages when compared with the SO,
e.g. high linearity since no signal-dependent on-resistance is presented, as well
as long-term reliability. Besides that, ROs can operate faster than SOs since no
turned-on transient is required. However, like SOs, ROs also disable the use of
many traditional SC structures and suffer, equally, from the front-end interfacing

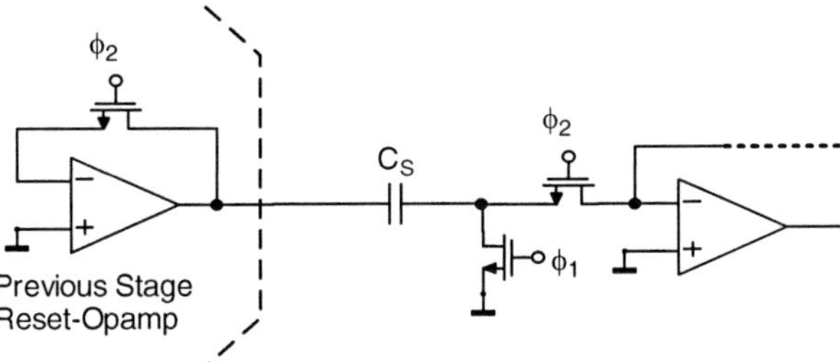

Fig. 2.14 Reset-opamp technique

problems. Unlike the operation of SO the ROs in reset phase can assure zero differential voltage, but the common-mode voltage is still undefined thus traditional CMFB techniques cannot be applied. As a result, current RO implementations are all in pseudo-differential mode (i.e. two singled-ended opamps instead of single fully-differential opamps) which imposes double power consumption.

2.5.6 *Switched-RC Techniques*

Both SO and RO circuits are truly compatible with future CMOS deep-submicron scaling since they both require no on-chip high-voltage to be successfully implemented. However, with these two techniques no floating switches can be used implying that the next-stage sampling capacitors are equivalently perpetually connected to the opamp's output nodes, which prevents the use of some optimum SC structures, like the fold-back type S/H and MDAC that have higher feedback factors. The requirement of switching-off or resetting the opamp to ground rather than to mid-supply voltage also imposes the need of common-mode level-shifting circuits.

Another technique that has been recently proposed is called switched-RC technique as shown in Fig. 2.15 [6] which is also truly a low-voltage technique. Actually, it uses switched-RC branches as a complete substitution of floating switches in such a way that the use of traditional optimum SC structures becomes possible. In this case, the sampling capacitor C_S is splitted into two $C_S/2$ capacitors, one connected to V_{DD} in the reset phase while the other is connected to GND. The function of floating switches will be replaced by two resistors R located in each split capacitor branches. During the sampling phase $\phi1$ both capacitors $C_S/2$ sample the previous opamp's output voltage through the resistors R. During the charge transfer phase the switches MSP and MSN are both turned-on forming a voltage divider with the resistor R to discharge the capacitors to the voltages near VDD and GND (depending on the voltage division ratio). This operation equivalently discharges the sampling capacitor to mid-supply without the use of floating switches.

The switched-RC technique presents similar advantages when compared with switched- and reset-opamps, namely it doesn't need on-chip high voltage and

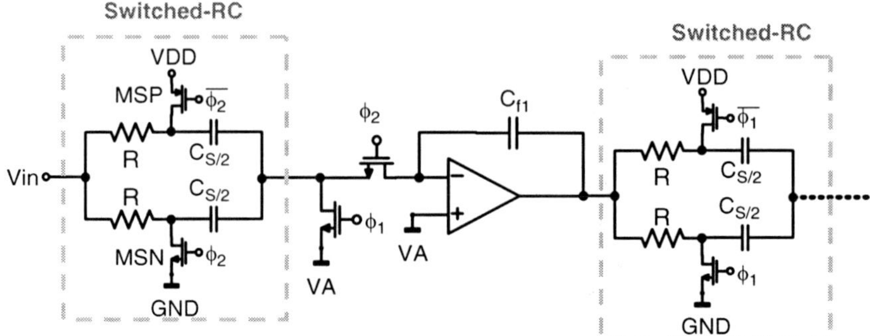

Fig. 2.15 Switched-RC technique

exhibits high linearity. Since it doesn't rely on specific switching action of previous stage's opamp no extra input sampling circuit is needed, allowing the use of traditional optimum SC circuits. In addition, no level shifter is needed since the split switched-RC branch cancelled the common-mode charges from upper and lower branches. However, since resistive dividers are present which are formed by R and on-resistances of MSP and MSN in phase 2, signal feedthrough will exist at the charge-transferring phase which can be only alleviated by choosing large R and large sizes of MSP/MSN. But, there is a trade-off in choosing the value of R since large R together with the output impedance of the opamp increases the sampling time-constant, which implies that switched-RC is not suitable for high-speed applications.

2.6   Summary

This chapter has presented the analysis and discussion of various types of typical low-voltage analog design challenges that lead to performance degradation compared with the traditional nominal-voltage implementations. Due to the reduced supply voltage headroom the signal swing must be reduced which leads to a reduction in SNR, or in other words the power consumption must be increased to maintain the system performance. Furthermore, short-channel effects become prominent as the technology scales down, degrading various intrinsic device parameters such as transconductance, output resistance and slowing down the trends of increasing speed as feature size reduces. Low supply voltage also leads to functionality problems of opamp and switches, although various solutions are available (with different types of trade-offs), such as the use of two-stage opamps, careful switch positioning, clock-boosting and bootstrapped switches, switched-opamp and reset-opamp techniques, as well as switched-RC techniques. Although different solutions exist to allow low-voltage circuits to be implemented, the performance degradation still cannot be avoided due to the unavailability of many useful circuit techniques, such as cascoding, the difficulty in designing CMFB circuits, front-end interface to continuous-time input signal, gain-and-offset compensation, and comparator designs. Considerations about these low-voltage problems will be further discussed in the next chapter.

References

1. J. Crols, M. Steyaert, Switched-opamp: an approach to realize full CMOS switched-capacitor circuits at very low power supply voltages. in *IEEE J. Solid-St. Circ.* **29**, 936–942 (Aug 1994)
2. M. Waltari, K.A.I. Halonen, 1-V 9-bit pipelined switched-opamp ADC. in *IEEE J. Solid-St. Circ.* **36**(1), 129–134 (Jan 2001)
3. M. Keskin, U. Moon, G.C. Temes, A 1-V 10-MHz clock-rate 13-bit CMOS $\Delta\Sigma$ modulator using unity-gain-reset op amps. in *IEEE J. Solid-St. Circ.* **37**(7), 817–824 (July 2002)
4. V.S.L. Cheung, H.C. Luong, W.H. Ki, A 1-V 10.7 MHz switched-opamp bandpass $\Sigma\Delta$ modulator using double-sampling finite-gain-compensation technique. in *IEEE J. Solid-St. Circ.* **37**(10), 1215–1225 (Oct 2002)
5. D. Chang, G. Ahn, U. Moon, Sub-1-V design techniques for high-linearity multistage pipelined analog-to-digital converters. in *IEEE T Circuits Syst I* **52**(1), 1–12 (Jan 2005)
6. G.-C. Ahn et al., A 0.6V 82dB 16 audio ADC using switched-RC integrators, in *Digest of Technical Papers of International Solid-State Circuits Conference (ISSCC)* (Feb 2005), pp. 166–167, 597
7. M. Keskin, A low-voltage CMOS switch with a novel clock boosting scheme. in *IEEE T Circuits Syst II* **52**(4), 185–188 (April 2005)
8. A.M. Abo et al., A 1.5 V, 10-bit, 14 MS/s CMOS pipeline analog-to-digital converter, in *Digest of Technical Papers of VLSI Circuits* (June 1998), pp. 166–169
9. J. Steensgaard, Bootstrapped low-voltage analog switches, in *Proceedings of IEEE International Symposium on Circuits and Systems (ISCAS)* (June 1999), pp. 29–32
10. M. Dessouky et al., Very low-voltage digital-audio $\Delta\Sigma$ modulator with 88-dB dynamic range using local switch bootstrapping. in *IEEE J. Solid-St. Circ.* **36**(3), 349–355 (March 2001)
11. H.-C. Kim, D.-K. Jeong, W. Kim, A 30mW 8b 200MS/s pipelined CMOS ADC using a switched-Opamp technique, in *Digest of Technical Papers of IEEE International Solid-State Circuits Conference (ISSCC)* (Feb 2005), pp. 284–285, 598
12. J. Li, G. Ahn, D. Chang, U. Moon, A 0.9-V 12-mW 5-MSPS algorithmic ADC with 77-dB SFDR. in *IEEE J. Solid-St. Circ.* **40**(4), 960–969 (April 2005)
13. D. Chang, U. Moon, A 1.4-V 10-bit 25-MS/s pipelined ADC using opamp-reset switching technique. in *IEEE J. Solid-St. Circ.* **38**(8), 1401–1404 (Aug 2003)
14. Intel, Moore's Law (Online, 2003), http://www.intel.com/research/silicon/mooreslaw.htm
15. C.G. Sodini, P.K. Ko, J.L. Moll, The effect of high fields on MOS device and circuit performance. in *IEEE T. Electron Dev.* **31**, 1386–1393 (Oct 1984)
16. P.K. Ko, in *Advanced MOS Device Physics*, ed. by N.G. Einspruch, G. Gildenblar (Academic, San Diego, CA, 1998)
17. A. Razavi, *Design of Analog CMOS Integrated Circuits* (McGraw-Hill, New York, 2001)
18. Y. Nakagome et al., Circuit techniques for 1.5-3.6-V battery-operated 64-Mb DRAM. in *IEEE J. Solid-St. Circ.* **26**(7), 1003–1010 (July 1991)
19. L. Singer et al., A 12b 65 MSample/s CMOS ADC with 82 dB SFDR at 120 MHz, in *Digest of Technical Papers of IEEE International Solid-State Circuits Conference (ISSCC)* (Feb 2000), pp. 38–39
20. D. Aksin et al., Switch bootstrapping for precise sampling beyond supply voltage. in *IEEE J. Solid-St. Circ.* **41**(8), 1938–1943 (Aug 2006)

21. S.K. Gupta, V. Fong, A 64-MHz clock-rate sigma-delta ADC with 88-dB SNDR and -105-dB IM3 distortion at a 1.5-MHz signal frequency. in *IEEE J Solid-St. Circ.* **37**(12), 1653–1661 (Dec 2002)
22. B. Hernes et al., A 1.2V 220MS/s 10b pipeline ADC implemented in 0.13/spl mu/m digital CMOS, in *Digest of Technical Papers of IEEE International Solid-State Circuits Conference (ISSCC)* (Feb 2004), pp. 256–257
23. V. Peluso et al., A 1.5-V-100-μW $\Delta\Sigma$ modulator with 12-b dynamic range using the switched-opamp technique. in *IEEE J. Solid-St. Circ.* **32**(7), 943–952 (July 1997)
24. A. Baschirotto et al., A 1-V 1.8-MHz CMOS switched-opamp SC filter with rail-to-rail output swing. in *IEEE J. Solid-St. Circ.* **32**(12), 1979–1986 (Dec 1997)
25. V. Peluso et al., A 900-mV low-power $\Delta\Sigma$ A/D converter with 77-dB dynamic range. in *IEEE J. Solid-St. Circ.* **33**(12), 1887–1997 (Dec 1998)
26. V. Cheung et al., A 1V CMOS switched-Opamp switched-capacitor pseudo-2-path filter, in *Digest of Technical Papers of IEEE International Solid-State Circuits Conference (ISSCC)* (Feb 2000), pp. 154–155
27. V.S.L. Cheung et al., A 1-V 10.7-MHz switched-opamp bandpass SD modulator using double-sampling finite-gain-compensation technique, in *Digest of Technical Papers of IEEE International Solid-State Circuits Conference (ISSCC)* (Feb 2001), pp. 52–53
28. P.Y. Wu, V.S.L. Cheung, H.C. Luong, A 1-V 100-MS/s 8-bit CMOS switched-opamp pipelined ADC using loading-free architecture. in *IEEE J. Solid-St. Circ.* **42**(4), 730–738 (April 2007)

Chapter 3
Advanced Low Voltage Circuit Techniques

3.1 Introduction

While both ROs and SOs can be utilized to avoid the floating switches at the opamp output nodes, many design challenges still exist in a low-voltage environment resulting in overall performance degradation when compared with a conventional design.

One of the main problems in low-voltage designs is the handling of the Common-Mode (CM) level. The CM level of the virtual ground of the opamp is always biased near the supply rail in low-voltage operation, while the opamp output CM level is usually biased at the middle of the supply rails to maximize the output swing. The difference between the virtual ground and output CM level requires an extra level-shifting circuit [1] which can degrade the feedback factor and thus the speed performance. Furthermore, the absence of floating switches still prevents the usage of many useful conventional circuit techniques. Firstly, the traditional CMFB circuit [2, 3] which is necessary in the implementation of fully-differential opamps cannot be applied. To overcome this problem well established SO techniques are available [4, 5], however for RO circuits there is no truly CMFB circuit available to allow fully-differential operation (for instance the CMFB circuits from [1, 6, 7] can only be applied in pseudo-differential circuits, i.e. two single-ended opamps instead of a fully-differential architecture). Secondly, both the RO and SO techniques rely on the previous stage's opamp to be reset or switched-off, respectively, which will create problems in the foremost front-end stage (directly coupling with the continuous-time input signal). As a result, traditional active solutions require extra front-end Track-and-Reset (T/R) stages [8–10] while a passive solution creates continuous-time signal feedthrough [5]. Finally, traditional gain-and-offset compensation techniques, such as Correlated Double Sampling (CDS) [11, 12] cannot be applied due to the unavailability of floating switches, imposing the utilization of two-stage opamps to achieve enough gain with low-speed in low-voltage designs. Reduced supply voltage also prevents the use of cascode current

S.-W. Sin et al., *Generalized Low-Voltage Circuit Techniques for Very High-Speed Time-Interleaved Analog-to-Digital Converters*, Analog Circuits and Signal Processing, DOI 10.1007/978-90-481-9710-1_3, © Springer Science+Business Media B.V. 2010

sources for accurate current biasing, making it difficult to optimize the performance of opamps under process variations.

In addition to the previous drawbacks, the limitation to utilize conventional circuits also places a large penalty on the overall performance (power or/and speed) of RO and SO circuits when compared with traditional designs, due to the impossibility of applying various useful special design techniques such as double-sampling [13] and opamp sharing [14], as well as the possible degradation of the feedback factor induced by architectural constraints.

This chapter serves two purposes: initially, it summarizes comprehensively the above several kinds of advanced low-voltage problems; finally, it proposes solutions for simultaneously achieving low-voltage, power efficiency and high-speed operation for both RO and SO circuits. The novel techniques here proposed have been generalized and they can be effectively applied to both RO and SO switched-capacitor circuits, including integrators in sigma-delta converters, Sample-and-Hold (S/H) and Multiplying Digital-to-Analog Converters (MDAC) used in pipelined ADCs.

3.2 Virtual-Ground Common-Mode Feedback and Output Common-Mode Error Correction

3.2.1 Low-Voltage CMFB Design Challenges

Designing CMFB for a fully-differential circuit constitutes a great challenge in a low-voltage environment. Figure 3.1 shows a traditional SC-CMFB implementation [2, 3] that is widely used in conventional SC circuits. In the presence of a large voltage swing in both V_{out+} and V_{out-}, the floating switches (inside the dashed circles) connected to these two output nodes cannot be easily turned on.

In the context of SO circuits [5, 15–17] the opamps are switched off in one phase to produce an high-impedance state in the opamp output, thus allowing it to be pulled up to a well-defined voltage (typically supply or ground potential, but other fixed potential can also be used) as shown in Fig. 3.2 [4, 5]. This allows resetting the capacitor C_1 in the SC-CMFB circuit and thus the CM voltage in the next phase can be controlled by those well-defined voltages in a similar way as in traditional

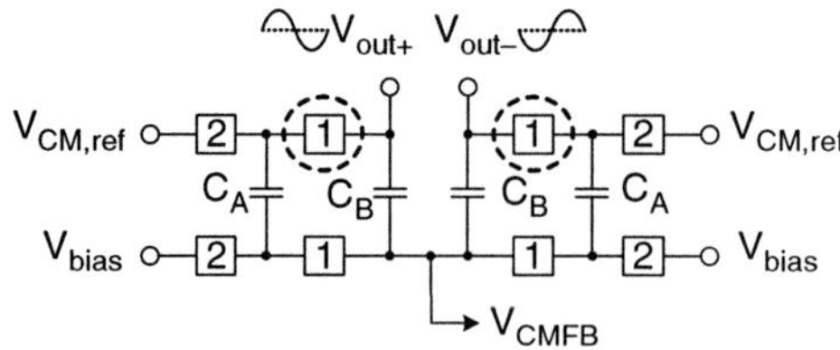

Fig. 3.1 Traditional SC-CMFB circuit

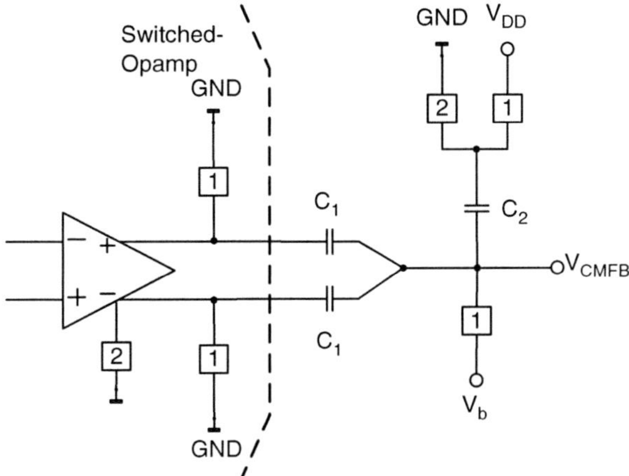

Fig. 3.2 SC-CMFB circuit used in SO [4, 5]

SC-CMFB techniques, then permitting fully-differential implementation of SO circuits.

The same principle cannot be applied to the RO architectures [1, 6, 18, 19] where the opamps are reset in an unity-gain configuration to discharge the next stage sampling capacitors. This operation produces a well defined 0-V differential mode resetting voltage, but the CM output at this phase is still undefined without the use of any CMFB. Due to the lack of such well-defined resetting voltage in the reset phase, the CMFB circuit from Fig. 3.2 cannot be utilized. Currently, all RO techniques employ pseudo-differential opamps [1, 6, 18, 19] (which actually are two single-ended opamps) as it is exemplified by the RO SC amplifier circuit (which can be used as an S/H or an MDAC) from the small figure in Fig. 3.3 in order to stabilize the opamp output CM voltage in both phases. However, it implies doubling the number of opamps as well as the corresponding power consumption. Furthermore, single-ended opamps are slower than their fully-differential counterparts due to the additional mirror pole created by the differential-to-single-ended converter.

3.2.2 Novel Virtual Ground CMFB Technique

Instead of directly controlling and stabilizing the output CM voltage, the same purpose can be achieved indirectly by controlling the virtual ground CM voltage [20–22], and this possibility will be addressed here first before a solution is presented. As shown in Fig. 3.3, the previous stage RO circuit resets at phase 2 to discharge the sampling capacitors C_1, while in phase 1 the stage resets itself to a level of $V_{G,CM}$ to discharge the next stage sampling capacitors, where $V_{G,CM} = (V_{G+} + V_{G-})/2$. It is assumed that the previous stage RO is similar to the current stage, also resetting at $V_{G,CM}$, and having an output CM voltage of $V_{in,CM}$ in amplification

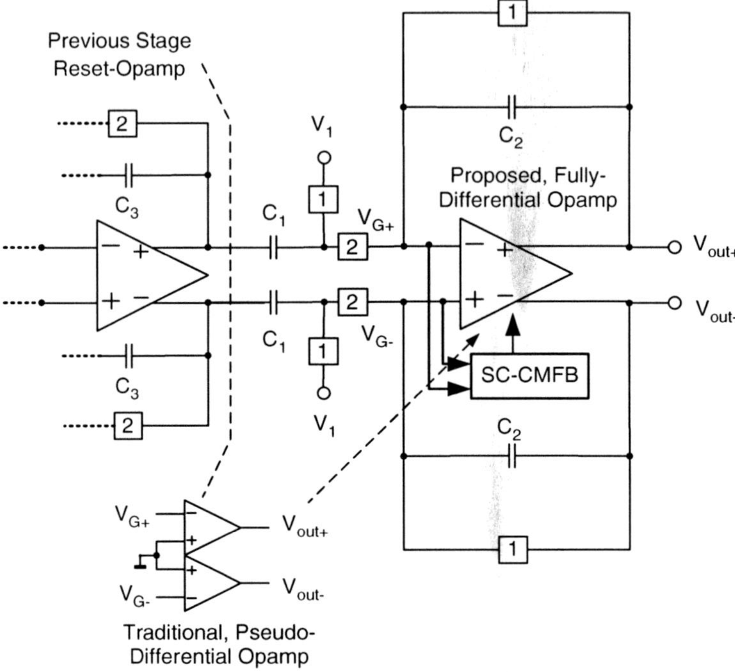

Fig. 3.3 A low-voltage SC amplifier circuit using fully-differential RO

mode (i.e. phase 1), which is actually the input CM voltage in the current stage. A simple mathematical relationship between various signal's CM level in phase 2 can be easily derived as follows:

$$V_{out,CM}[\phi 2] = \frac{C_1}{C_2} V_{in,CM} + V_{G,CM} - \frac{C_1}{C_2} V_1 \tag{3.1}$$

where V_1 is a known, fixed potential that can be used to achieve the level-shifting function (Please see Section 3.4 for detailed analysis). Assuming $V_{in,CM}$ has a known voltage value (as in the usual case which is set by the previous stage), Eq. (3.1) possesses two unknown variables $V_{out,CM}[\phi 2]$ and $V_{G,CM}$, i.e. only one constraint equation derived from the external SC networks. If a CMFB circuit (implying another constraint equation) is applied such that either $V_{out,CM}[\phi 2]$ (the traditional output CMFB) or $V_{G,CM}$ (the proposed Virtual Ground CMFB, VG-CMFB) are defined, then the other unknown variable can also be determined. This simple equation clearly shows how the control of the virtual ground CM voltage can lead to the stabilization of opamps' output CM level, and in this way, the pseudo-differential opamp-pair in Fig. 3.3 can be replaced by one fully-differential opamp, thus saving half of the power consumption. Notice also that the target value of the $V_{G,CM}$ should be set with a voltage difference in the order of 0.1–0.3 V from the supply rails, such that the opamp's output transistors always remain in the saturation region during the reset phase for high-speed operation. If biasing of $V_{G,CM}$ to VDD or ground is necessary, then an SC floating battery must be utilized [1].

Compared with the traditional output CMFB control, the proposed method can be easily implemented in a low-voltage environment because: (a) the signal swing in the virtual ground is much smaller than the one in the output of the opamps due to the opamp's differential gain, thus alleviating the floating switch problem; (b) the output CM level is usually different in the two phases of low-voltage circuits and thus also difficult to be controlled, while the virtual ground is usually fixed in both phases to allow a suitable biasing of the input differential pair. However, stabilizing the output CM by controlling the virtual ground CM level is not as accurate as the traditional output CMFB method, since any inaccuracy in the virtual ground CM voltage will be amplified by $1/\beta$ to the output CM voltage (with β as the feedback factor), and the CM charge injection error will also appear, both leading to output CM errors. Furthermore, these type of CM errors can be easily accumulated (e.g. in the integrating capacitors of SC integrators or pipelining stages). Since the CM output voltage is not required to be so accurate as the differential mode signal the proposed technique can be utilized alone in low-speed (with small switch sizes leading to smaller charge injection errors) or in small number of pipelining stages (where the CM error accumulation does not saturate the last-stage's opamp). While in high-speed circuits the CM error accumulations become significant (due to larger switches), as well as in integrators, this technique can be combined with the novel Output CM Error Correction (O-CMEC), as discussed later, to achieve an accurate output CM error control.

3.2.3 Practical Implementation of VG-CMFB

Figure 3.4 shows the proposed VG-CMFB technique [20–22], including the differential pair in the main opamp and the reference generation circuit. In addition to the normal transconductance operation, the differential pair of an opamp can also serve as a virtual ground CM voltage detector, requiring no extra CM detector circuits. As the CM bias currents of M1A and M1B are set by the tail current source M0 of the differential pair, then $V_{GS1A,CM} = V_{GS1B,CM} = V_{GS1,CM}$ are kept unchanged and the differential pair reproduces the shifted version of the virtual ground CM voltage at $V_{tail} = V_{G,CM} - V_{GS1,CM}$. The reference generation circuit is a scaled-down version of the differential pair thus simulating its biasing condition and also reproducing $V_{tail,ref} = V_{G,ref} - V_{GSB1}$, in which $V_{G,ref}$ represents the desired virtual ground potential. Since the reference is generated and the virtual ground CM level has also been detected and represented through V_{tail}, conventional SC-CMFB circuit techniques can be employed. V_{CMFB} has a similar meaning as in conventional SC-CMFB which could be applied to the gate of one pair of current source transistors in the main opamp, providing adjustment of the output CM level, and also through the connection of an external SC network to the virtual ground CM

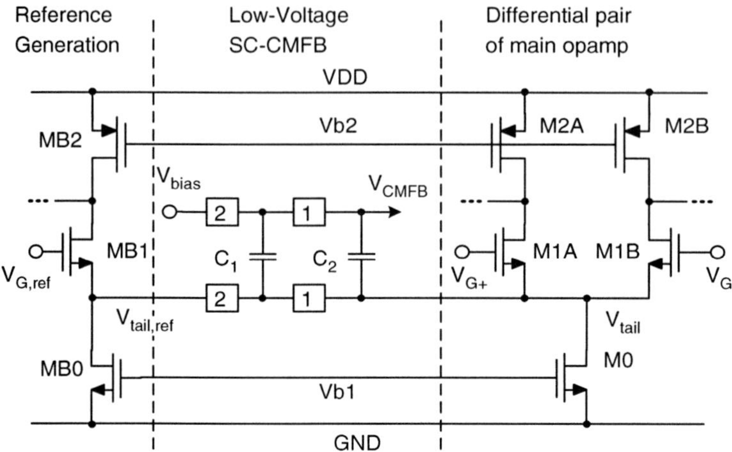

Fig. 3.4 The proposed virtual ground CMFB technique

potential. V_{bias} is the nominal bias value of V_{CMFB}. Also, all the switches in Fig. 3.4 can be easily turned on since no signal swings are presented in any nodes of the SC-CMFB circuit (thus can be biased near the supply rails).

3.2.4 Output Common-Mode Error Correction

As discussed previously, the VG-CMFB technique controls the virtual-ground CM potential and through the external passive network stabilizes the output CM level. Then, any errors entering the external passive network (e.g. charge injection) can affect the output CM voltages, and the errors can be easily accumulated in the integrators or pipelining stages. When these errors become relevant the VG-CMFB technique can be enhanced further through the combination with a novel Output Common-Mode Error Correction (O-CMEC) technique, also proposed here and shown in Fig. 3.5 [22]. This circuit is an improved version of the pseudo-differential implementation from [1]. If the output CM voltage is at the desired value, then the net charge Δq injected to V_{CMFB} node by the O-CMEC will be cancelled out, while any output CM deviation will imply a correction charge by the negative CM feedback loop. For example, if a CM error is injected into the external feedback network, and correspondingly, the output CM voltage of the opamp, in phase 2 of Fig. 3.3, is being pulled down from the designed value (normally the mid-supply). Then, referring to Fig. 3.5, $V_{out,CM}$ will be pulled down, disturbing the equilibrium and resulting in a negative charge Δq injected into the V_{CMFB} node. This action will pull down the node voltage V_{CMFB} leading the output CM voltage to be pulled up by the CM gain (since the CMFB must be designed as negative feedback loop). Appendix A will describe in detail the operating principle of the VG-CMFB + O−CMEC

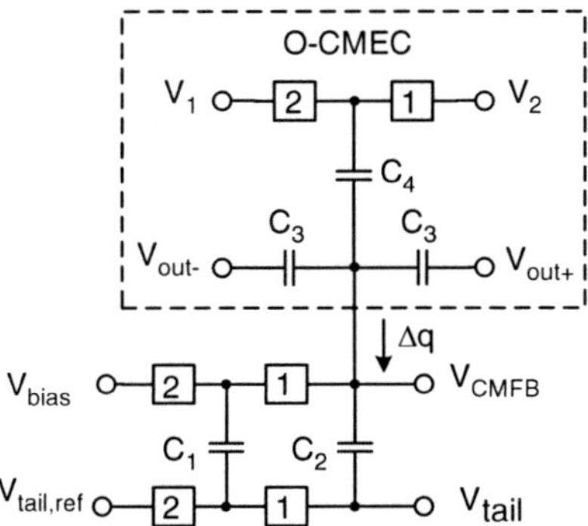

Fig. 3.5 The proposed output CM error correction circuit

technique and presents also the derivation of the following design equation for selection of the capacitor values:

$$V_{out,CM}[\phi 1] - V_{out,CM}[\phi 2] = \frac{C_4}{2C_3}(V_2 - V_1) \tag{3.2}$$

meaning that the capacitor ratio depends on the difference of the output CM voltage in the two phases and also the reference voltages V_1 and V_2 that can be either VDD or ground.

Compared with the implementation from [1], the proposed technique offers the following improvements: (a) combined with the VG-CMFB circuit it is designed for fully-differential opamps, instead of being only applied to the pseudo-differential mode as in [17]; (b) only one correction circuit is required here, while two circuits were required in the previous implementation [17]; (c) the O-CMEC injects charges into V_{CMFB} node which is an internal node of the CMFB loop, and this only reduces the common-mode feedback factor and does not affect the differential-mode feedback factor. From [1] the injection point is the virtual ground of the main opamp which will degrade both feedback factors thus leading to speed penalties. However, the O-CMEC circuit cannot be used independently since the VG-CMFB imposes the near-steady-state operating point of the output CM level first and only after that the output CM error can be corrected by O-CMEC.

3.3 Cross-Coupled Passive Sampling Interface

3.3.1 Problems in Existing Solutions

Both RO and SO circuits rely on the precedent stage opamps, which simulate floating switches, to discharge the sampling capacitors. This creates a difficulty in the foremost front-end stage that directly interfaces with the input continuous-time signal. As a result, extra efforts must be devoted to design the front-end input sampling circuits. Several input interfaces have been already proposed before in both RO and SO circuits [5, 8–10, 18] and one of such from [10] is presented in Fig. 3.6a. In the figure a resistor is utilized as voltage divider during the resetting phase, thus allowing the discharge of the sampling capacitor. This passive circuit cannot be imbedded into normal SC front-end building blocks (e.g. S/H) because this interface actually simulates the normal non-inverting, non-delay switched-capacitor branch (Fig. 3.6b) and, when combined with the opamp it can only provide a T/R output for the continuous-time input signal of Fig. 3.7 (and thus being an extra T/R stage, in addition to the S/H that normally is required). Similar techniques are also used in [8, 9, 18] where the input interface is actually an active inverting resistive amplifier, again implementing only the T/R function. An extra T/R stage not only consumes one extra front-end opamp's power, but it will also place severe limitations on the noise, linearity and speed, since the performance of such T/R is directly linked to the full resolution (or dynamic range) of the whole system.

Passive sampling interfaces also exist [5, 19] that simulate the usual inverting, half-delay SC branches, thus allowing truly S/H/R waveforms (as shown in

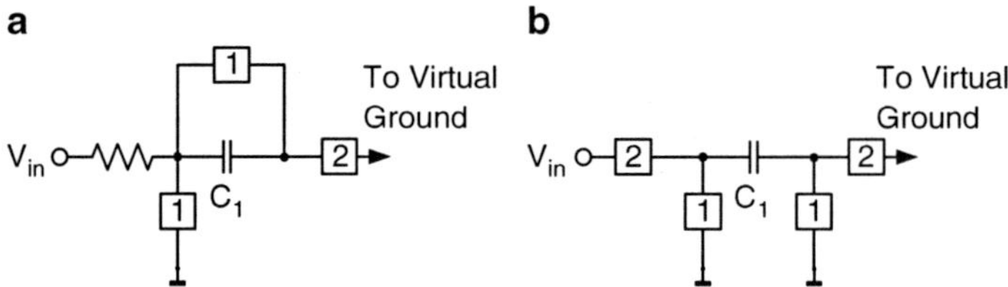

Fig. 3.6 (**a**) Low-voltage passive sampling circuit [10]; (**b**) a non-inverting, non-delay SC branch

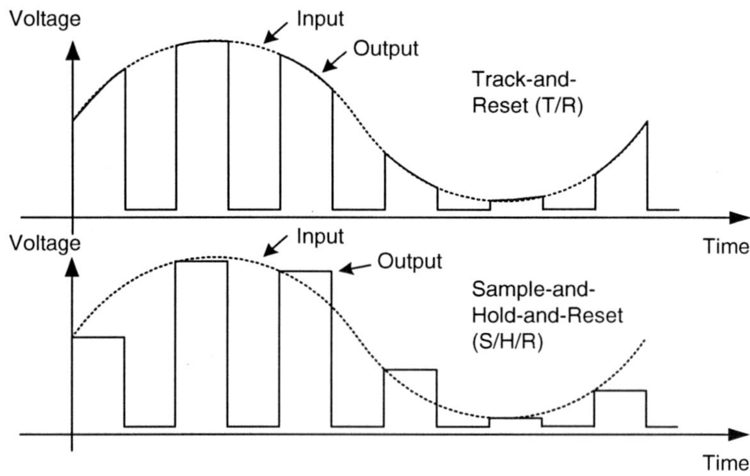

Fig. 3.7 Track-and-Reset (T/R) versus Sample-and-Hold-and-Reset (S/H/R)

Fig. 3.7) and then it can be imbedded into normal front-end SC building blocks. However, extra clock phases are needed [5] and most importantly, direct signal feedthrough from continuous-time input to output can occur during the amplification phase [5, 19]. To suppress such signal feedthrough some limitations are also placed on the size selection of various component values including switches and resistors. For example, to suppress the signal-feedthrough, the value of the resistor must be large to form a high-ratio resistive divider with the switch on-resistance, and this will place large limitations on the sampling time constant and degrades the operating speed [19].

3.3.2   Cross-Coupled Passive Sampling Interface

Figure 3.8 shows the proposed low voltage cross-coupled passive sampling circuits (in differential configuration) [21–23] as a remedy for the drawbacks previously discussed. When compared with the sampling circuit from Fig. 3.6, this circuit is equivalent to the common inverting, half-delay SC branch. The resistor R (similarly as in Fig. 3.6) provides resistive division with the on-resistance of switches S1 and it will also allow the discharge operation of the sampling capacitor C_1 in phase $\phi 2$. An extra pair of capacitors C_2 is added, each with the same size of C_1. The circuit operates as follows: In phase $\phi 1$, the differential signal will be sampled by the capacitor pair C_1, while the capacitor pair C_2 is reset; In phase $\phi 2$, the switches S1 are turned on and form a voltage divider with the resistors R, thus the signal swing on nodes V_{x+}, V_{x-} will be attenuated and all the switches connected to this nodes can now be easily turned on. In this phase the capacitor pair C_1 is discharged between the virtual ground of the opamps and the attenuated signal, causing direct signal feedthrough interference in the charge transferring phase. The capacitor pair C_2 is

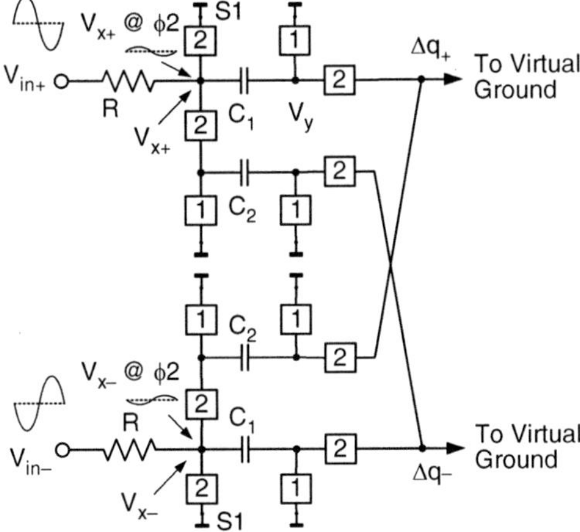

Fig. 3.8 The proposed low-voltage cross-coupled passive sampling circuits

now used to eliminate it since C_2 is connected between the virtual ground and the attenuated signal in the opposite differential path. If $C_1 = C_2$, then the signal feedthrough will be cancelled and only the differential charge sampled in C_1, in phase $\phi 1$, is transferred to the virtual ground.

A simple quantitative analysis of the differential charge transfer during phase $\phi 2$ in the circuit from Fig. 3.8 yields:

$$\Delta q_+ - \Delta q_- = -C_1 V_{in}[\phi 1] - \frac{R_{on}}{R + R_{on}}(C_2 - C_1)V_{in}[\phi 2] \tag{3.3}$$

where $\Delta q_+ - \Delta q_-$ represents the amount of differential charge transferred to the virtual ground at the end of $\phi 2$, R_{on} the on resistance of the switches S1, $V_{in} = V_{in+} - V_{in-}$ the differential input signal, and $V_{in}[\phi 1]$ the input signal at the end of $\phi 1$. From (3.3) it can be observed that the differential signal feedthrough will appear during $\phi 2$, but it will be canceled if $C_1 = C_2$ through a cross-coupling action of the passive sampling branch. To gain a deeper insight on the effectiveness of the feedthrough cancellation mechanism, Fig. 3.9 shows a plot of the direct signal-feedthrough attenuation (in dB) versus the resistive divider ratio on R_{on}/R and the mismatch in C_1 and C_2, compared with the conventional techniques (equivalent to $C_2 = 0$) [19]. This plot shows, for example, that even with moderate resistive divider ratio (e.g. 0.1) and capacitor matching (1%), the proposed circuit still provides high signal feedthrough attenuation (-60 dB) compared with the conventional structure (-20 dB only). To further demonstrate the effectiveness of the circuits in Fig. 3.8, Table 3.1 shows the performance comparison of the S/H circuit shown in Fig. 3.10 with and without the crossed-coupled capacitor C_2. The crossed-coupled passive sampling circuit shows a superior signal feedthrough attenuation with the trade-off of reduced speed and a little-bit increased circuit noise (Assume 1.2 Vpp differential signal swing), due to the main reason of the extra capacitor and the reduced feedback factor.

The on resistance R_{on} of switches S1 depends on the gate-source voltage and then on V_{x+} and V_{x-}, modulating the voltage on those nodes by voltage division

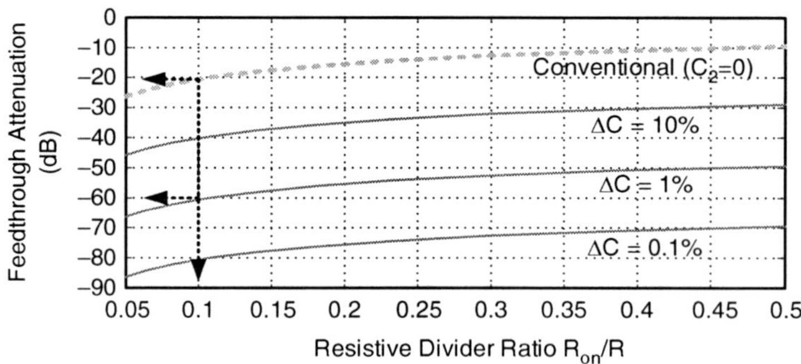

Fig. 3.9 A plot of signal-feedthrough attenuation versus R_{on} and capacitor mismatch in C_1 and C_2

Table 3.1 Performance comparison of Fig. 3.8 with and without C_2

Case	Target resolution	Value of C_2	Feedback factor	Sampling capacitor	Capacitor matching[a]	Normalized speed	Normalized thermal noise	Signal to noise (dB)	Signal to feedthrough[b] (dB)	Signal to (Feedthrough + Noise) (dB)
1	10	$C_2 = C_1$	0.33	500 fF	0.3%	8	12	62	70	61.5
2	10	$C_2 = 0$	0.5	500 fF	0.3%	12	8	64	20	20
3	12	$C_2 = C_1$	0.33	4 pF[c]	0.1%	1	1.5	71	80	70.6
4	12	$C_2 = 0$	0.5	4 pF[c]	0.1%	1.5	1	73	20	20

[a]Extracted from the foundry datasheet

[b]Extracted from Fig. 3.9 with $R_{on}/R = 0.1$

[c]For 2-b increased resolution the capacitor sized should be increased by a factor of 16, however 4 pF is chosen here to reduce the power consumption

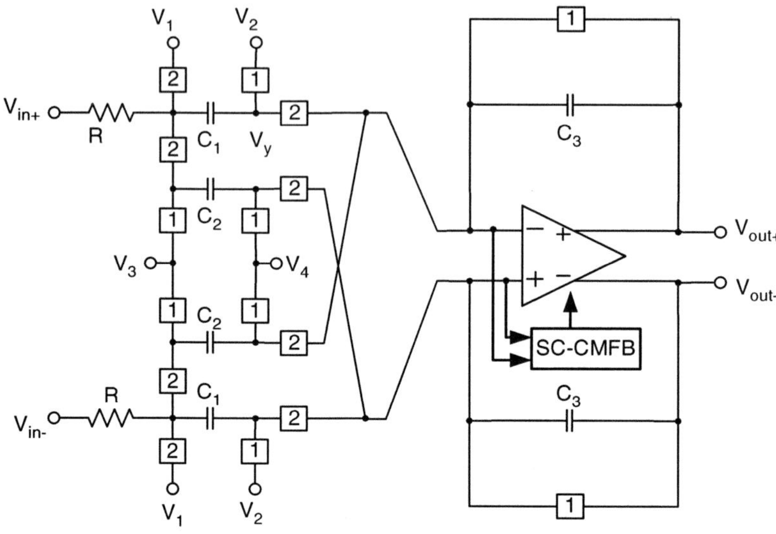

Fig. 3.10 The front-end S/H with crossed-coupling passive interface

which causes harmonic distortion in phase 2. However, the proposed circuit is insensitive to such distortion because the produced even harmonics will be suppressed by fully-differential operation, while those odd harmonics are canceled by the action of the cross-coupled sampling branch, since they are similar to the attenuated signal feedthrough, and, only the differential input signal charges sampled in C_1 during phase 1 are transferred. In addition, the proposed circuit only needs simple two-phase clocking, and also the signal feedthrough will be canceled provided that C_1 matches C_2 and the size of switches S1 are chosen such that S1 can be turned on. This greatly relaxes the limitations of component values selection found before in [5, 19]. Also, since the input signal is sampled in phase $\phi1$ while charge transfer is performed in phase $\phi2$, the sampling circuit simulates the normal inverting, half-delay SC branches which can provide S/H/R output with reset- or switched-opamp, since the input continuous-time signal is decoupled in the charge transferring phase. This implies that the circuit can be imbedded in normal SC building blocks such as S/H and integrators, eliminating the need of additional T/R stages and thus reducing power consumption. Although extra passive components (capacitors C_2) are added, the proposed circuit consumes lesser area and power compared to other implementations since the added passive component is used to trade with the whole extra T/R stages that contain more passive components as well as one extra opamp. A drawback of the proposed circuit is its reduced feedback factor which would impose speed limitations, when compared with previous designs, because one additional pair of capacitors is connected to the opamp virtual ground.

3.4 Voltage-Controlled Level Shifting

Handling CM voltage in low-voltage designs is not as simple as in traditional circuits since the CM voltage at the virtual ground of the opamp is usually biased near VDD or ground for the proper operation of the input differential-pair, while the CM voltage at the opamp's output is usually at the mid-supply to maximize the output swing. Also, due to the nature of RO and SO circuits the output CM level between two phases can be quite different. Such CM level difference originates the need for level-shifting circuits [1] which can be implemented by an SC branch connected to the virtual ground of the opamp to provide a constant DC charge every clock cycle. However, this method brings along a speed penalty as an SC branch connected to the virtual ground of the opamp degrades the feedback factor.

To avoid such disadvantage a novel technique is proposed here that can be designated as Voltage-Controlled Level-Shifting (VCLS) [22], which achieves the required level shifting through simple charge-redistribution principles. The operation principle can be demonstrated by considering again the RO S/H from Fig. 3.3 together with (3.1) that presented the dependence of the output CM voltage with the input CM level, virtual ground CM level and the fixed potential V_1. Actually the last term in (3.1) already inherently provides the required level shifting function through the controlling voltage V_1, thus not requiring the extra level-shifting SC branch and allowing higher speed of operation. For example in Fig. 3.3 with $C_1 = C_2$, if we choose $V_{out,CM}[\phi 2] = V_{in,CM} = 0.6$ V under $V_{DD} = 1.2$ V, and $V_{G,CM} = 0.9$ V for proper operation of opamp with NMOS differential pair, then V_1 should be chosen as $V_1 = 0.9$ V from (3.1).

Sometimes this technique cannot be directly applicable with some specific circuit configurations simply due to the improper value of V_1 in a low-voltage environment, such as in the case of $V_1 > V_{DD}$, $V_1 < GND$ or V_1 near the mid-supply leading to the floating switch difficulty. For example, considering Fig. 3.3 again with a gain-of-4 SC amplifier that will imply $C_1 = 4C_2$, and in this case V_1 would be calculated as 0.675 V, which is at around middle of the 1.2 V supply. In this case, the circuit can be rearranged to include more level-shifting controlling potentials, as shown in Fig. 3.11. The functionality of this circuit is exactly the same as the one in Fig. 3.3, except that an extra level-shifting controlling voltage V_2 has been added. The output CM voltage can now be evaluated as follows:

$$V_{out,CM}[\phi 2] = \frac{C_1}{C_2} V_{in,CM} + 2V_{G,CM} - \left(V_2 + \frac{C_1}{C_2} V_1 \right) \tag{3.4}$$

The last term of the equation corresponds to the level-shifting term and for the previous case with $C_1 = 4C_2$ it would be possible to choose $V_1 = 0.9$ V and $V_2 = 0$ V to achieve the required level-shift. Moreover, $V_{G,CM}$ can also be adjusted to contribute to the level-shift, if necessary.

The proposed VCLS technique achieves the level-shifting function without any extra SC branch connected to the virtual ground, and even no modification of the

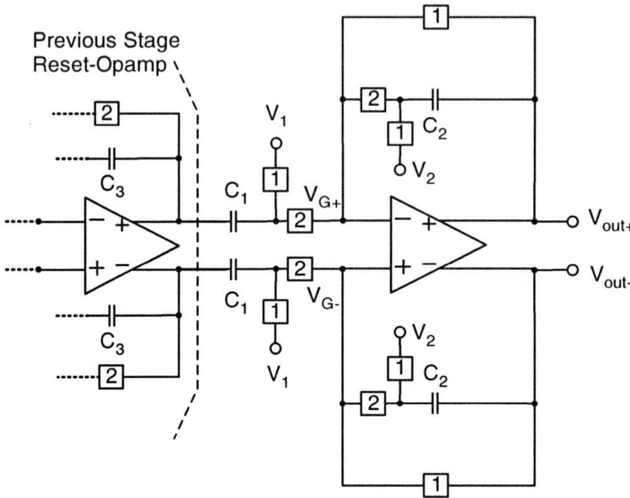

Fig. 3.11 An alternative for RO amplifier circuit

original circuit is necessary except the change of the fixed potentials, then avoiding speed penalty when compared with the circuit from [1]. The drawback of this technique is the requirement for extra controlling voltages. However, in most applications the fixed potentials can be originated on the supply rails, and normally only one or two extra controlling voltages are required, as shown in the previous example, $V_1 = V_{G,CM} = 0.9$ V implying that they can be obtained from the same voltage buffer. On the other hand, high accuracy of those controlling voltages is also not required since they only affect the CM output voltage.

3.5 Feedback Current Biasing Technique

Figure 3.12 shows the current-mirror opamp with the proposed Feedback Current Biasing (FBCB) technique, which can also be applied in the auxiliary differential-difference amplifier discussed in next section. For correct operation the bias current of M3A is provided by currents' subtraction between M2A and M1A. The bias current of M1A is provided by the tail current source M0, which can have larger channel length to achieve accurate current matching over process variations (because it is not in the signal path). However, the channel length of M2A can't be too large, since the drain junction capacitance of M2A is in the main signal path and the phase margin will be degraded by this parasitic capacitance. This capacitance can be comparable to the gate capacitance of M3A since M2A is a PMOS transistor with large current handling capability. On the other hand, using a smaller channel length in M2A will imply large current spread over process corners due to the channel length modulation effect from the variation of the gate-source voltage

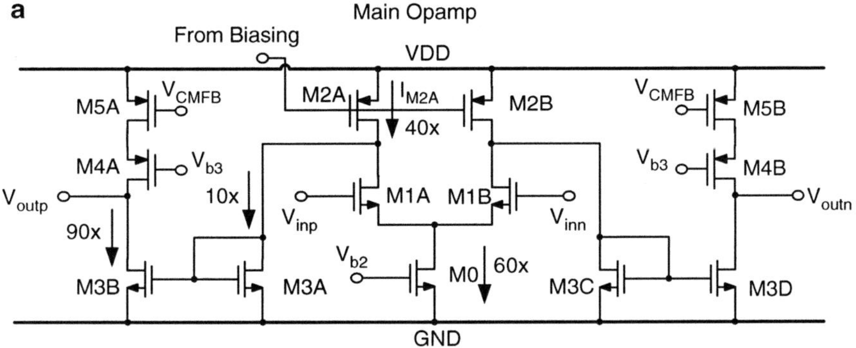

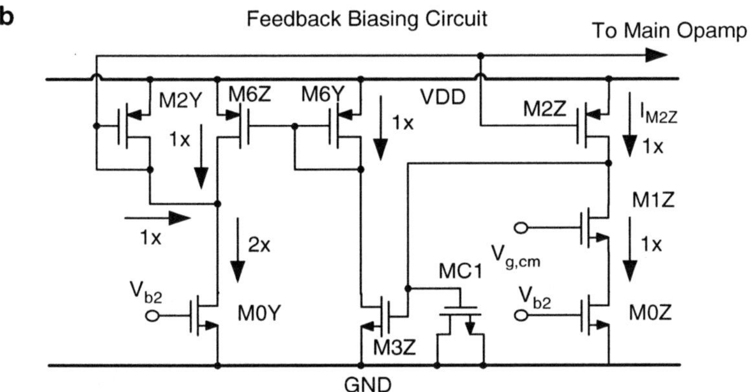

Fig. 3.12 A current mirror opamp (**a**) with feedback current biasing circuit (**b**)

V_{GS} in M3A. Since now the current in M3A would be the subtraction of M2A from M1A it will impose a very large current variation in M3A, as well as the output-stage transistor M3B which could affect the slew rate of the opamp.

This difficulty can be overcome by the proposed biasing circuit in Fig. 3.12. The biasing circuit will simulate the operating point of the main opamps, i.e. MxZ, MxY are the corresponding scaled down versions of Mx, MxA–MxD in the main opamp (e.g. M3Z is a scaled-down version of M3A to M3D). Figure 3.12 also exhibits the current relationship allowing a better understanding on how to set the current in the various branches, as well as the W/L ratio of various transistors. The feedback loop composed by M3Z, M6Y, M6Z, M2Y and M2Z ensures closely-tracked drain voltages in M2Z and M2A. Since their drain voltages are identical, the current matching between them is accurate over process variations even with small transistor lengths. On the other hand, choosing larger channel lengths for M0Z and M0 also allows accurate matching between them. Now, since the currents through M2Z and M0Z are identical, this technique guarantees closely tracked current-ratio between M2A and M1A, then minimizing M3A's current variation under small channel length (0.25 μm in this design).

To ensure that the feedback loop is negative, an additional current mirror pair M6Y and M6Z has been added. This feedback loop is a unity-gain feedback and the gain-bandwidth product of the loop gain can be calculated as:

$$GBW = \frac{(W/L)_{M6Z}}{(W/L)_{M6Y}} \cdot \frac{(W/L)_{M2Z}}{(W/L)_{M2Y}} g_{m,M3Z} / (C_{gs,M3Z} + C_{gs,MC1}) \tag{3.5}$$

considering that the drain parasitics are not dominant in the biasing circuit. In addition, there are two non-dominant poles:

$$p_1 = g_{m,M2Y} / (C_{gs,M2Y} + C_{gs,M2Z} + C_{gs,M2A} + C_{gs,M2B}) \tag{3.6}$$

$$p_2 = g_{m,M6Y} / (C_{gs,M6Y} + C_{gs,M6Z}) \tag{3.7}$$

Due to the large gate capacitance of the transistors M2A and M2B in the main opamp, $p_1 << p_2$ and the first non-dominant pole p1 severely degrades the phase margin of the loop. A large NMOS capacitor MC1 is thus added at the gate of M3Z to stabilize the feedback loop.

To verify the effectiveness of the technique, simulations have been done for the auxiliary differential-difference amplifier used in the first MDAC. AC analysis shows that the biasing loop achieves an open-loop gain of 27 dB, GBW of 5.4 MHz and phase-margin of 80°. Moreover, corner simulations are performed and the results are shown in three cases, in Table 3.2. For the first two cases, M2A and M2B are biased using conventional simple current mirror (without cascode), these two cases clearly show the trade-off between the output-stage current (and thus slew-rate) variations and the phase-margin through the selection of channel length of M2A/B. Using longer channel length can ensure lesser output-stage current variations (17–3.6%), with the cost of the reduced phase margin in the main opamp (61–46°). The last case shows that the proposed technique can achieve simultaneously improved phase margin (61°) and reduced process variations (1.6% only, even better than case 2).

Table 3.2 Comparison of conventional biasing scheme over the feedback current biasing technique

	Case 1	Case 2	Case 3
	Conventional biasing		Proposed
Channel length of M2A/B	0.25 μm	0.75 μm	0.25 μm
Output-stage current variation	±17%	±3.6%	±1.6%
Gain-bandwidth product	1.35 GHz	1.46 GHz	1.3 GHz
Phase margin @ $\beta = 1/5$	61°	46°	61°

3.6 Low-Voltage Finite-Gain-Compensation

3.6.1 The Need for Finite-Gain Compensation

In low-voltage designs the opamps are mainly restricted to traditional two-stage architectures [1, 20] (due to the impossibility of using cascode devices), which by their nature exhibit lower speed (due to the additional high-impedance node that needs miller compensation) and higher power consumption (more current branches). On the other hand, using single-stage opamps with low gain can cause serious systematic errors like nonlinearities in MDACs of pipelined ADCs or poles and zeros deviations in SC filters or sigma-delta modulators, unless the produced finite-gain error can be compensated, with, for example, traditional Correlated Double Sampling (CDS) techniques [11, 12] and the buffer compensation technique [24]. However, they cannot be applied in a low-voltage environment due to (a) limitations caused by floating switches problems and (b) by the fact that the opamp is switched-off or reset in one clock phase, which implies that it would not be idle and cannot be used to compensate the gain error. To overcome these drawbacks, a low-voltage gain compensation technique was addressed before in [17] but it has also a restriction of narrow-band operation (typically a bandpass sigma-delta modulator) that imposes a limitation to the signal band which must be located only and narrowly at $f_s/4$.

A Low-Voltage Finite-Gain Compensation (LV-FGC) technique is here proposed for wide-bandwidth and high-speed circuits [22, 25], including in detail the investigation of the feedback factor mismatch effect. To illustrate the idea behind the LV-FGC technique it would be necessary to consider a low-voltage reset-opamp amplifier (which can be used to model the MDAC used in a pipelined ADC), as shown in Fig. 3.13. Here, the upper part includes the usual main amplifier and only a single-ended version is shown for simplicity, although the real imple-mentation is fully-differential. In phase 1, the input signal from the previous stage is sampled in C_{s1}, while in phase 2 the previous stage's opamp resets to discharge the sampling capacitor C_{s1} to virtual ground. Again V_1 is a fixed potential that allow level-shifting which does not affect the differential signal being processed and then are not considered here (in Section 3.4 their calculation to achieve the required level-shift had been addressed). Without considering the proposed auxiliary ampli-fier, it can be derived that the amplifier implements the following arithmetic function (considering only differential information signals and including the effects of input parasitic capacitance C_{P1} and finite gain A_1):

$$V_{o1} = \frac{C_{s1}}{C_{f1}} V_{in} + \frac{1}{\beta_1}\left(-\frac{V_{o1}}{A_1}\right) \tag{3.8}$$

where

$$\beta_1 = \frac{C_{f1}}{C_{s1} + C_{f1} + C_{P1}} \tag{3.9}$$

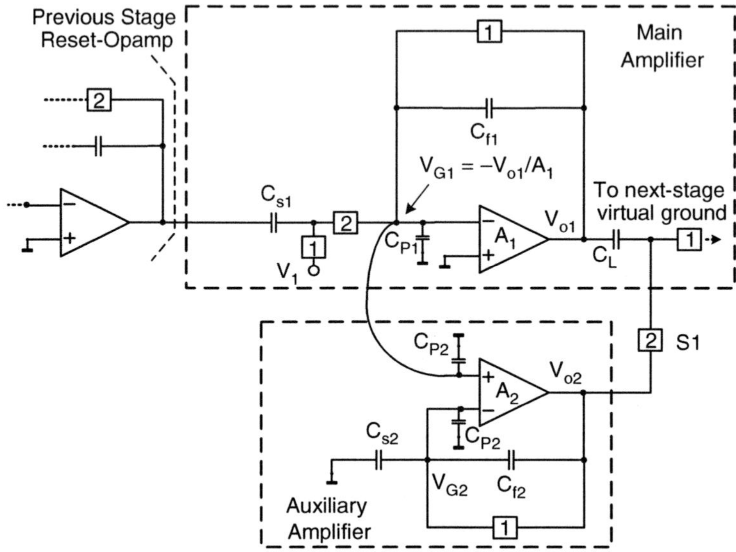

Fig. 3.13 Low-voltage gain-compensated SC amplifier

represents the feedback factor. In the right-hand side of (3.8) the V_{o1} term will not be combined with V_{o1} in the left-hand side because the term

$$- V_{o1}/A_1 = V_{G1} \tag{3.10}$$

actually corresponds to the virtual ground voltage of A_1 and it is clearly evident now that the virtual ground error voltage is being amplified by the inverse of the feedback factor $(1/\beta_1)$ to the output, leading to a finite-gain error. As illustrated in Fig. 3.13, the fundamental idea of the proposed solution is to sense the main opamp virtual ground voltage by an auxiliary amplifier in non-inverting configuration, amplify it by the same feedback factor, and then feed it into the bottom plate of C_L to cancel the gain error.

The output of the auxiliary amplifier can be derived as:

$$V_{o2} = \frac{C_{s2} + C_{f2} + C_{P2}}{C_{f2}} V_{G2} = \frac{1}{\beta_2} V_{G2} \tag{3.11}$$

where

$$V_{G2} = V_{G1} - V_{o2}/A_2 \tag{3.12}$$

which means that the gain of the auxiliary amplifier will also contribute to the total gain error. Substituting (3.12) into (3.11) yields

$$V_{o2} = \frac{1}{\beta_2} V_{G1} - \frac{V_{o2}}{\beta_2 A_2} = \frac{V_{G1}}{\beta_2[1 + 1/(\beta_2 A_2)]} = -\frac{V_{o1}}{\beta_2 A_1[1 + 1/(\beta_2 A_2)]} \tag{3.13}$$

Now for the main amplifier, the feedback factor should be modified in order to account for the input parasitic C_{P2} from the auxiliary amplifier:

$$\beta_1' = \frac{C_{f1}}{C_{s1} + C_{f1} + C_{P1} + C_{P2}} \qquad (3.14)$$

Then the equivalent output voltage that is sampled into the capacitor C_L in phase 2 (Fig. 3.13) is

$$V_{o1} - V_{o2} = \frac{C_{s1}}{C_{f1}} V_{in} + \text{Gain error} \qquad (3.15)$$

where the gain error can be evaluated as

$$\text{Gain error} = \frac{1}{\beta_1'} \left(-\frac{V_{o1}}{A_1} \right) - \frac{-V_{o1}}{\beta_2 A_1 [1 + 1/(\beta_2 A_2)]}$$

$$= \left(-\frac{V_{o1}}{A_1} \right) \frac{\beta_2 - \beta_1' + 1/A_2}{\beta_1' \beta_2 [1 + 1/(\beta_2 A_2)]}$$

$$\approx \left(-\frac{V_{o1}}{A_1 A_2} \right) \frac{A_2(\beta_2 - \beta_1') + 1}{\beta_1' \beta_2} \qquad (3.16)$$

with the assumption of $\beta_2 A_2 \gg 1$. Also, supposing that a mismatch exists between β_1' and β_2 with $\beta_1' - \beta_2 = \Delta\beta$, $(\beta_1' + \beta_2)/2 = \beta$, $\beta_1' = \beta + \Delta\beta/2$ and $\beta_2 = \beta - \Delta\beta/2$, it implies that (3.16) can be simplified to:

$$\text{Gain error} = \left(-\frac{V_{o1}}{A_1 A_2} \right) \frac{1 + A_2 \Delta\beta}{\beta^2 \left[1 - \left(\frac{\Delta\beta}{2\beta} \right)^2 \right]} \approx \frac{1}{\beta} \left(-\frac{V_{o1}}{A_1 A_2} \right) \left(\frac{1}{\beta} + A_2 \frac{\Delta\beta}{\beta} \right)$$

$$\approx \frac{1}{\beta} \left(-\frac{V_{o1}}{A_1 A_2} \right) \left(\frac{1}{\beta} + A_2 \frac{\Delta\beta}{\beta} \right) \qquad (3.17)$$

assuming $\Delta\beta/\beta << 1$ and $\beta A_1 >> 1$ such that V_{o2} is small when compared with V_{o1} as indicated by (3.13). Comparing (3.17) with (3.8) it can be deducted that the effective gain can be expressed as:

$$\text{Effective gain} = \frac{A_1 A_2}{\frac{1}{\beta} + A_2 \frac{\Delta\beta}{\beta}} \qquad (3.18)$$

If there is no mismatch between the two feedback factors, then $\Delta\beta = 0$ and the effective gain would have been boosted from A_1 to $\beta A_1 A_2$ by the proposed technique.

To achieve the proposed gain compensation, the feedback factor of both amplifiers should be matched as:

$$\frac{C_{s1} + C_{P1} + C_{P2}}{C_{f1}} = \frac{C_{s2} + C_{P2}}{C_{f2}} \tag{3.19}$$

Note that C_{P1} and C_{P2} depend on the biasing condition of the differential pairs of both amplifiers. However, Eq. (3.19) can be easily satisfied by using the same sizes and biasing conditions for both of the input differential pairs (that makes $C_{P1} = C_{P2}$) and by choosing the following capacitor ratios:

$$C_{s1} = 2C_{s2}, \; C_{f1} = 2C_{f2}, \tag{3.20}$$

Alternatively, several choices satisfying (3.19) can be used, like for example a scaled-down version of the differential pair in the auxiliary amplifier, with the choice of the corresponding capacitor ratios also according to (3.19).

As indicated by (3.18), the mismatch between the feedback factors of the main and auxiliary amplifiers reduces the effective gain of the proposed technique. But, as it will be shown next, the mismatch-induced gain-reduction is not significant in the normal range of feedback factor mismatch values. A plot of the normalized effective gain (normalized to $\beta A_1 A_2$) versus $\Delta\beta/\beta$ is shown in Fig. 3.14 with $\beta = 0.23$ including the simulation result with a real opamp (described in detail in Section 3.6.2) that is provided here for verification purposes. The plot shows that the effective gain will be reduced as the feedback factor mismatch increases. For example, if we choose both of the opamp gains as 49 dB, the effective gain reduces from $\beta A_1 A_2 = 85$ dB with no mismatch to 81 dB for $\Delta\beta/\beta = 1\%$. Even for such large mismatch the proposed scheme still provides satisfactory performance compared with the one without gain-compensation. The reason for such insensitivity derives from the fact that the auxiliary amplifier is processing the main opamp gain error, as stated in (3.13), which has a relatively small magnitude in the order of a

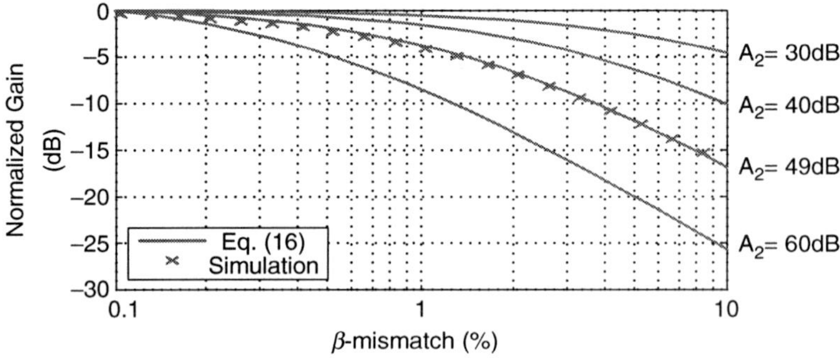

Fig. 3.14 A plot of normalized effective gain versus β mismatch

few mV (for instance 10 mV), and the β mismatch (e.g. 1%) only causes negligible error (100 μV). In practice, the parasitics capacitance associated with the switches that are connected to virtual grounds also contributes to mismatch in the feedback factor, and if this becomes a problem, dummy switches can be used to improve such type of matching. Also, the amount of gain reduction will also increase as the auxiliary amplifier gain increases due to the fact that as the effective gain error reduces (since the gain increases), the magnitude of the total gain error is reduced and thus more sensitive to the mismatch of the feedback factor.

In Fig. 3.13 a switch S1 is used in the auxiliary opamp output to disconnect C_L at phase 1, such that C_L can be discharged to the next stage virtual ground. This switch can be turned on and off without any problem since the swing in the output of the auxiliary amplifier is quite small and can be set near the supply rails. Similarly, the nonlinearity produced by its on-resistance is also negligible.

Compared with the implementation presented in [24] which also utilizes the auxiliary amplifier to sense the error voltage from the virtual-ground of the opamp, with the following disadvantages (the readers are referred to [24] for detailed discussion): (a) the implementation in [24] cannot be used in low-voltage environment due to the floating switches problem; (b) in [24] the error signal needs to be fed into the subsequent stages for correction of the gain-error, which either leads to feedback factor degradation and thus to speed penalties (due to extra SC-branch injected to the next-stage virtual-ground) or to the requirement of extra "shadow" pipelined stages, adders, as well as analog delays which will increase the power consumption. The proposed scheme corrects the gain-error immediately in the current stage, avoiding extra injection to the virtual-ground and thus optimizing both the speed and the power consumption.

3.6.2 *Auxiliary Differential-Difference Amplifier*

It might seem that the proposed technique will significantly increase the power consumption since an additional amplifier is needed. However, the additional power consumption can be traded-off with the significant increase of the single-stage opamps' Gain-Bandwidth Product (GBW). Moreover, fully-differential operation can be utilized by applying the previously described low-voltage VG-CMFB circuit, which can further cut half of the opamps' power when compared to pseudo-differential designs.

The implementation of this technique requires the use of a non-inverting amplifier, as shown in Fig. 3.13. The traditional implementation of a fully-differential non-inverting auxiliary amplifier is not possible, since both opamp inputs are used as virtual grounds in the differential circuit. Instead, the Differential-Difference Amplifier (DDA) can be used to implement the fully-differential non-inverting amplifier [26]. Figure 3.15 presents the Auxiliary Differential-Difference Amplifier (A-DDA) current-mirror opamps that will be used in the MDAC which will also be used in the real chip prototype of a pipelined ADC later. The A-DDA consists

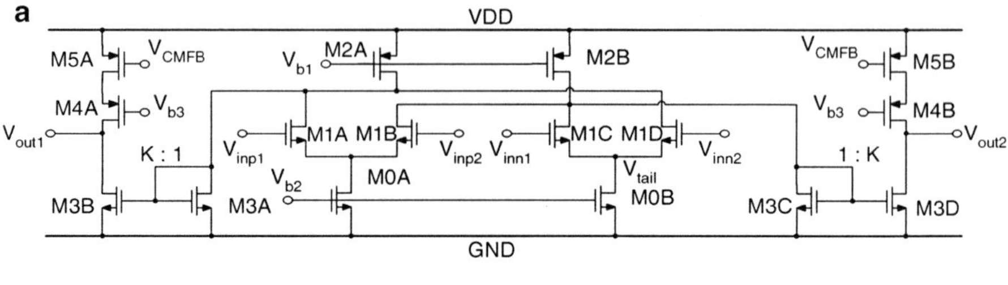

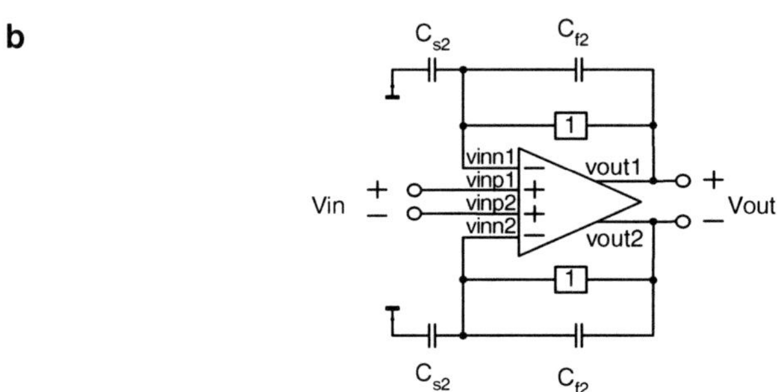

Fig. 3.15 Auxiliary Differential-Difference Amplifier topologies: (a) circuit schematic and (b) the symbol

of two differential pairs with four inputs (two for virtual grounds and two for non-inverting inputs), with their currents sum at the drain of M2A and M2B, afterwards folded into the diodes M3A and M3C and later mirrored through M3B and M3D to the outputs. The output voltage and GBW of the opamp can be expressed as [17, 26]:

$$V_{out1} - V_{out2} = Kg_{m1}[(V_{inp1} - V_{inp2}) - (V_{inn1} - V_{inn2})]R_{out} \qquad (3.21)$$

$$GBW = Kg_{m1}/(2\pi C_{Ltot}) \qquad (3.22)$$

where R_{out} represents the equivalent output resistance, K is the current mirror ratio and C_{Ltot} is the total capacitive load. In Fig. 3.15a the VG-CMFB circuit can be applied from V_{tail} in the right-side differential-pair to VCMFB to stabilize the CM voltage in the auxiliary amplifier, and the virtual ground CM voltage of the other differential-pair will be set by the main opamp's virtual ground CM level (See Fig. 3.13).

Several special techniques are used in the A-DDA of Fig. 3.15 to further improve its speed, namely: (a) NMOS differential pairs have inherently larger transconductance than their PMOS counterparts and their drain current is folded into the diodes M3A and M3C, which are also NMOS such that the phase margin is not degraded significantly by this current mirror pole in the signal path. Such configuration can achieve potentially higher speed than the traditional current mirror opamp with

NMOS differential pair and PMOS current mirror; (b) Cascode transistors M4A and M4B are added into the output current branch to shield the PMOS current source transistors M5A and M5B from the Miller multiplication of the output node, which can significantly increase the input capacitance of the CMFB feedback point and thus slow down the common-mode response. Also, due to the cascode, it is possible to use minimum length transistors in M5A and M5B to further reduce their input parasitics. The cascode transistor may approach the vicinity of triode region in a low-voltage environment, but since the output resistance is dominated by the NMOS side (M3B and M3D), the resulting nonlinearity will be suppressed. Such arrangement can still provide a 6 dB benefit to the gain as the output resistance now becomes $r_{o,M3B}$ rather than $r_{o,M3B}//r_{o,M5A}$.

3.7 Low-Voltage Offset-Compensation

In time-interleaved ADCs offset-mismatch among various channels creates fixed-offset tones at multiples of f_s/M, with f_s as the overall sampling rate and M as the number of channels [27]. In order to achieve high dynamic range of such system, the offset must be compensated. In this section, the low-voltage offset compensation techniques will be proposed which can be embedded into the previous discussed crossed-coupled S/H in Fig. 3.10 as well as the gain-compensated MDAC in Fig. 3.13.

3.7.1 The Crossed-Coupled S/H

The S/H with a low-voltage offset compensation technique is proposed as shown in Fig. 3.16 with its single-ended equivalent circuits shown in Fig. 3.17 (for both phases). To obtain an S/H gain of 1 it would be necessary to have $C_1 = C_2 = C_3 = C$. In phase 1 (Fig. 3.17a) C_2 is connected between the virtual grounds (V_{g+} and V_{g-} in Fig. 3.3) in the reset mode, thus equivalently sampling $2V_{OS}$ into C_2, where V_{OS} is the opamp offset voltage. A charge conservation equation (differential) can be written in phase 2 (Fig. 3.17b):

$$C_1\left[(V_{in}[\phi 1] - 0) - \left(\frac{R_{on}}{R + R_{on}}V_{in}[\phi 2] - V_{OS}\right)\right]$$
$$+ C_2\left[(-V_{OS} - V_{OS}) - \left(-\frac{R_{on}}{R + R_{on}}V_{in}[\phi 2] - V_{OS}\right)\right] \quad (3.23)$$
$$+ C_3[0 - (V_{out}[\phi 2] - V_{OS})] = 0$$

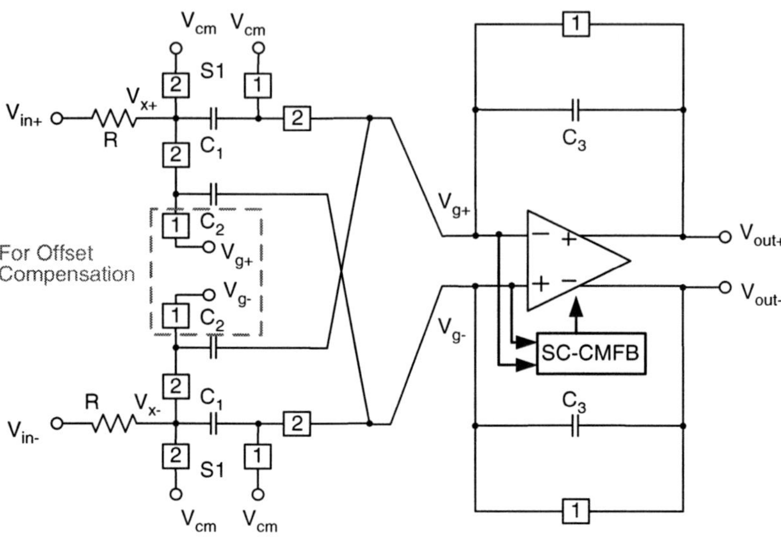

Fig. 3.16 Offset-compensated crossed-coupled S/H

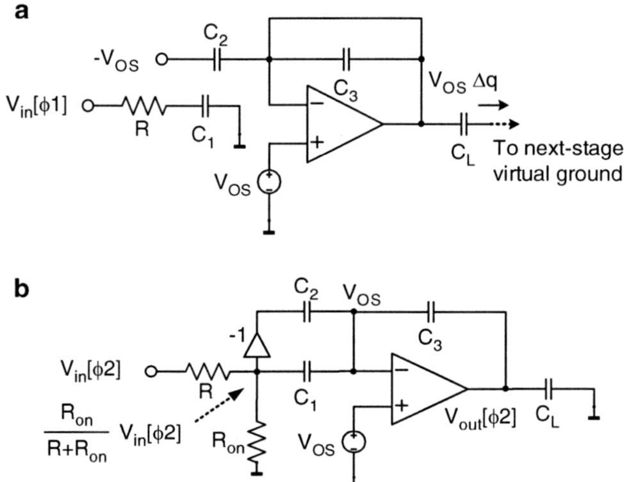

Fig. 3.17 Single-ended equivalent circuit of Fig. 3.16 in (a) phase 1 and (b) phase 2

leading to

$$V_{out}[\phi 2] = V_{in}[\phi 1] + V_{OS} \tag{3.24}$$

However, as derived from the operation of the reset-opamp circuit, the S/H will be reset (to V_{OS}) in phase 1, as illustrated in Fig. 3.17a, implying that the charge package discharged to the next-stage virtual ground can be expressed as:

$$\Delta q = C_L\{(V_{in}[\phi 1] + V_{OS}) - V_{OS}\} = C_L V_{in}[\phi 1] \tag{3.25}$$

showing that the opamp offset voltage is compensated without the use of any floating switches.

3.7.2 The SC Amplifier

In addition to the gain-compensation also an offset-compensation scheme is proposed which can be used in the SC amplifier in Fig. 3.13 to suppress the offset error generated simultaneously by the main and auxiliary amplifiers (as shown in Fig. 3.18). Notice the changes of switches connection in Fig. 3.18 as compared to Fig. 3.13 to enable the offset compensation. If V_{OS1} and V_{OS2} are the offset of the main and auxiliary amplifiers, respectively, then the main amplifier holds (considering only differential voltages):

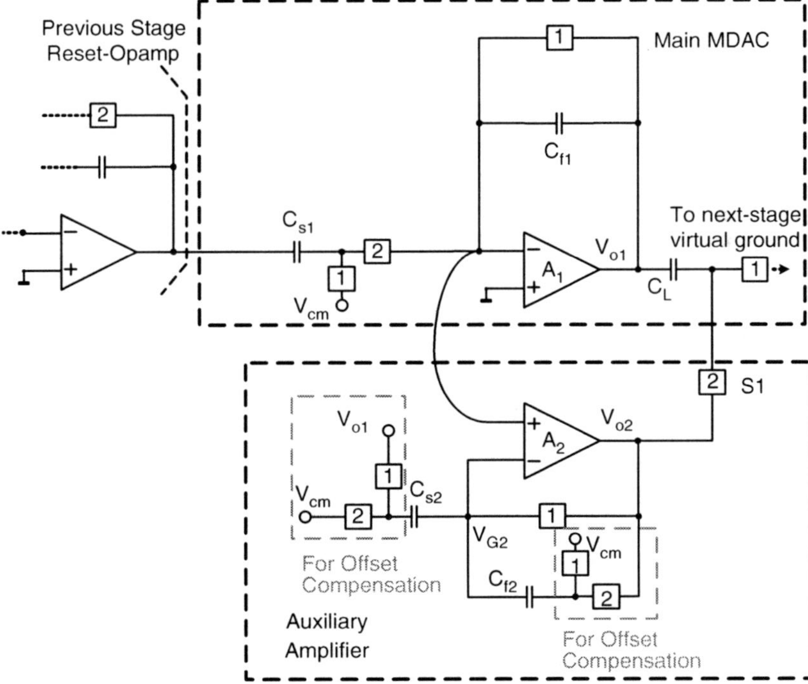

Fig. 3.18 Gain- and offset-compensated SC amplifier

$$C_{s1}[(V_{in}[\phi 1] - 0) - (0 - V_{OS1})] + C_{f1}[0 - (V_{o1}[\phi 2] - V_{OS1})] = 0 \qquad (3.26)$$

that leads to

$$V_{o1}[\phi 2] = \frac{C_{s1}}{C_{f1}} V_{in}[\phi 1] + \frac{C_{s1} + C_{f1}}{C_{f1}} V_{OS1} \qquad (3.27)$$

Besides, the auxiliary amplifier will process both offsets from the main and the auxiliary amplifiers as follows (with the voltage at the negative terminal of A_2 as $V_{OS1} + V_{OS2}$):

$$C_{s2}\{[(V_{OS1} - (V_{OS1} + V_{OS2})] - [0 - (V_{OS1} + V_{OS2})]\}$$
$$+ C_{f2}\{[(0 - (V_{OS1} + V_{OS2})] - [V_{o2}[\phi 2] - (V_{OS1} + V_{OS2})]\} = 0 \qquad (3.28)$$

which can be represented by

$$V_{o2}[\phi 2] = \frac{C_{s2}}{C_{f2}} V_{OS1} \qquad (3.29)$$

Notice that the following relationship is used in Section 3.6 for ensuring matched feedback factors in both amplifiers for gain compensation:

$$\frac{C_{s1}}{C_{f1}} = \frac{C_{s2}}{C_{f2}} \qquad (3.30)$$

Finally, by combining (3.27), (3.29) and (3.30), the voltage stored in the capacitor C_L during phase 2 can be expressed as:

$$(V_{o1} - V_{o2})[\phi 2] = \frac{C_{s1}}{C_{f1}} V_{in}[\phi 1] + V_{OS1} \qquad (3.31)$$

Again the main amplifier will be reset to V_{OS1} in phase 1 to discharge C_L resulting in an offset-free charge transferred to the next stage.

3.8　Summary

This chapter presented several low-voltage circuit techniques summarized together with their advantages in Table 3.3. They serve as comprehensive solutions to various problems, like CMFB, power-hungry front-end T/R, level-shifting, increased process sensitivity and finite gain and offset errors, in low-voltage, high-speed, power efficient Reset-Opamp and Switched-Opamp circuits. They comprise the Virtual-Ground CMFB (VG-CMFB) combined with Output

Table 3.3 Summary of the low-voltage techniques

Low-voltage techniques	Main improvements in low-voltage designs
Virtual-Ground CMFB with Output CM Error Correction (VG-CMFB + O-CMEC)	Power reduction (two pseudo-diff. opamp to one fully-diff. opamp)
Crossed-Coupled Passive Sampling Interface (CCPSI)	Power reduction (eliminate the extra T/H) Accuracy (Low signal-feedthrough error)
Voltage-Controlled Level Shifting (VCLS)	Speed (increased feedback factor)
Feedback Current Biasing (FBCB)	Reduce opamp's sensitivity to process variations
Low-Voltage Finite-Gain Compensation (LV-FGC)	Improved opamp's speed/power trade-offs (allow the use of high-speed low-gain single-stage opamp)
Low-Voltage Offset Compensation (LV-OC)	Canceling opamp's offset for time-interleaved operation

Common-Mode Error Correction (O-CMEC), Crossed-Coupled Passive Sampling Interface (CCPSI), Voltage-Controlled Level-Shifting (VCLS), Feedback Current Biasing (FCB), Low-Voltage Finite-Gain Compensation (LV-FGC) and Low-Voltage Offset Compensation techniques (LV-OC).

References

1. M. Keskin, U. Moon, G.C. Temes, A 1-V 10-MHz clock-rate 13-bit CMOS $\Delta\Sigma$ modulator using unity-gain-reset op amps. in *IEEE J. Solid-St. Circ.* **37**(7), 817–824 (July 2002)
2. R. Castello, P.R. Gray, A high-performance micropower switched-capacitor filter. in *IEEE J. Solid-St. Circ.* **SC-20**(6), 1122–1132 (Dec 1987)
3. O. Choksi, L.R. Carley, Analysis of switched-capacitor common-mode feedback circuit. in *IEEE Trans. Circuits Syst. II* **50**(12), 906–917 (Dec 2003)
4. M. Waltari, K. Halonen, A switched-opamp with fast common mode feedback, in *Proceedings of International Conference on Electronics, Circuits and Systems (ICECS)*, vol. 3 (Sept 1999), pp. 1523–1525
5. M. Waltari, K.A.I. Halonen, 1-V 9-bit pipelined switched-opamp ADC. in *IEEE J. Solid-St. Circ.* **36**(1), 129–134 (Jan 2001)
6. D. Chang, U. Moon, A 1.4-V 10-bit 25-MS/s pipelined ADC using opamp-reset switching technique. in *IEEE J. Solid-St. Circ.* **38**(8), 1401–1404 (Aug 2003)
7. L. Wu, M. Keskin, U. Moon, G. Temes, Efficient common-mode feedback circuits for pseudo-differential switched-capacitor stages, in *Proceedings of International Symposium on Circuits and Systems (ISCAS)*, vol. 5 (May 2000), pp. 445–448
8. A. Baschirotto, R. Castello, G.P. Montagna, Active series switch for switched-opamp circuits. in *IEE Electron. Lett.* **34**(14), 1365–1366 (July 1998)
9. S. Karthikeyan, A. Tamminneedi, C. Boecker, E.K.F. Lee, Design of low-voltage front-end interface for switched-op amp circuits. in *IEEE Trans. Circuits Syst. II* **48**(7), 722–726 (July 2001)
10. M. Keskin, A novel low-voltage switched-capacitor input branch. in *IEEE Trans. Circuits Syst. II* **50**(6), 315–317 (June 2003)

11. C.C. Enz, G.C. Temes, Circuit techniques for reducing the effects of op-amp imperfections: auto-zeroing, correlated double-sampling, and chopper stabilization. P in *IEEE* **84**(11), 1584–1614 (Nov 1996)
12. U. Seng-Pan, R.P. Martins, J.E. Franca, Highly accurate mismatch-free SC delay circuits with reduced finite gain and offset sensitivity, in *Proceedings of IEEE International Symposium on Circuits and Systems (ISCAS)*, vol. 2 (May 1999), pp. 57–60
13. J.F.F. Rijns, H. Wallinga, Spectral analysis of double-sampling switched-capacitor filters. in *IEEE Trans. Circuits Syst.* **38**(11), 1269–1279 (Nov 1991)
14. B.-M. Min, P. Kim, F.W. Bowman III, D.M. Boisvert, A.J. Aude, A 69-mW 10-bit 80-MSample/s pipelined CMOS ADC. in *IEEE J. Solid-St. Circ.* **38**(12), 2031–2039 (Dec 2003)
15. J. Crols, M. Steyaert, Switched-opamp: an approach to realize full CMOS switched-capacitor circuits at very low power supply voltages. in *IEEE J. Solid-St. Circ.* **29**, 936–942 (Aug 1994)
16. P.Y. Wu, V.S.L. Cheung, H.C. Luong, A 1-V 100-MS/s 8-bit CMOS switched-opamp pipelined ADC using loading-free architecture. in *IEEE J. Solid-St. Circ.* **42**(4), 730–738 (April 2007)
17. V.S.L. Cheung, H.C. Luong, W.H. Ki, A 1-V 10.7 MHz switched-opamp bandpass $\Sigma\Delta$ modulator using double-sampling finite-gain-compensation technique. in *IEEE J. Solid-St. Circ.* **37**(10), 1215–1225 (Oct 2002)
18. D. Chang, G. Ahn, U. Moon, Sub-1-V design techniques for high-linearity multistage pipelined analog-to-digital converters. in *IEEE Trans. Circuits Syst. I* **52**(1), 1–12 (Jan 2005)
19. G.-C. Ahn et al., A 0.6V 82dB $\Delta\Sigma$ audio ADC using switched-RC integrators, in *Digest of International Solid-State Circuits Conference (ISSCC)* (Feb 2005), pp. 166–167, 597
20. S.-W. Sin, U. Seng-Pan, R.P. Martins, A novel very low-voltage SC-CMFB technique for fully-differential reset-opamp circuits, in *Proceedings of IEEE International Symposium on Circuits and Systems (ISCAS)* (May 2005), pp. 1581–1584
21. S.-W. Sin, U. Seng-Pan, Martins R.P, Novel low-voltage circuit techniques for fully-differential reset- and switched-opamps, in *Proceedings of PRIME'2005*, vol. 2 (July 2005), pp. 398–401
22. S.-W. Sin, U. Seng-Pan, R.P. Martins, Generalized circuit techniques for low-voltage high-speed reset- and switched-opamps. in *IEEE Trans. Circuits Syst. I* **55**(8), 2188–2201 (Sept 2008)
23. S.-W. Sin, U. Seng-Pan, R.P. Martins, A novel low-voltage cross-coupled passive sampling branch for reset- and switched-opamp circuits, in *Proceedings of IEEE International Symposium on Circuits and Systems (ISCAS)* (May 2005), pp. 1585–1588
24. A.M.A. Ali, K. Nagaraj, Background calibration of operational amplifier gain error in pipelined A/D converters. in *IEEE Trans. Circuits Syst. II* **50**(8), 631–634 (Sept 2003)
25. S.-W. Sin, U. Seng-Pan, R.P. Martins, A novel low-voltage finite-gain compensation technique for high-speed reset- and switched-opamp circuits, in *Proceedings of 2006 IEEE International Symposium on Circuits and Systems - ISCAS'2006* (May 2006), pp. 3794–3797
26. H. Alzaher, M. Ismai, A CMOS fully balanced differential difference amplifier and its applications. in *IEEE Trans. Circuits Syst. II* **48**(6), 614–620 (June 2001)
27. M. Gustavsson et al., *CMOS Data Converters for Communications* (Kluwer, Boston, MA, 2000)

Chapter 4
Time-Interleaving: Multiplying the Speed of the ADC

4.1 Introduction

While the current minimum feature size of IC fabrication technology limits the maximum achievable speed of electronic devices, parallel or time-interleaved (TI) architectures are one of the most effective solutions to boost the maximum speed of analog interfaces at the system level [1–7]. In principle the operating speed of the TI-ADC can grow linearly by increasing the number of parallel ADC channels, however various types of mismatches among different channels create modulation tones which degrade the performance of the TI-ADC, like offset [8, 9], gain [8, 9], timing [5, 8–11] as well as bandwidth mismatches (from the sampling RC time-constant) [9, 12–14]. Those mismatch effects must be fully characterized to achieve satisfactory performance in the design of TI-ADCs.

This chapter will present first the principle of TI-ADC followed by the analysis of various types of channel mismatches and their effects into the performance of the whole ADC. Closed-form SNDR for all types of mismatch errors will be derived statistically and generalized with the inclusion of channel transfer functions' shaping effects to the mismatch-induced fixed-pattern tones and modulation sidebands. More details on the bandwidth mismatch (which is more difficult to analyze due to its dynamic nature) will be addressed also, since they will degrade the performance of the TI-ADC system even at low input-frequencies due to its phase-mismatch component.

4.2 Time-Interleaved ADC Architecture

Figure 4.1a shows an example of TI-ADC architecture, which utilizes M identical sub-ADCs arranged in parallel. The sub-ADC in each time-interleaved path operates at a sampling frequency of f_s/M, where f_s denotes the overall equivalent sampling frequency of the whole system. Figure 4.1b shows the clocking waveforms for

S.-W. Sin et al., *Generalized Low-Voltage Circuit Techniques for Very High-Speed Time-Interleaved Analog-to-Digital Converters*, Analog Circuits and Signal Processing, DOI 10.1007/978-90-481-9710-1_4, © Springer Science+Business Media B.V. 2010

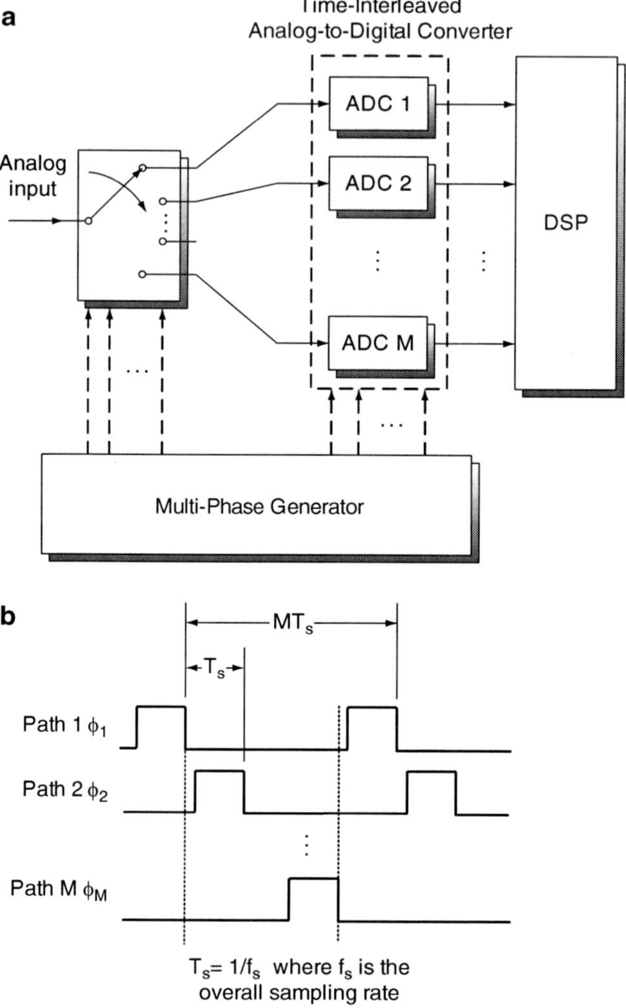

Fig. 4.1 A time-interleaved analog-to-digital converter. (**a**) The overall system architecture; (**b**) timing diagram

such parallel architecture. The sampling clocks applied to each path are delayed respectively by one overall sampling period $T = 1/f_s$, thus each subsystem samples the analog input signals in an alternative manner. By employing parallel architectures the effective sampling rate can be multiplied by the number (M) of TI paths. A special case of TI structure utilizes double-sampling techniques [1, 15], which are extensively used in SC circuits with an equivalent TI path number $M = 2$.

Although this parallel architecture seems very attractive in boosting the maximum operating speed of sampled-data systems, various component mismatches among different interleaved channels (which are theoretically identical) can seriously affect their performance [16], thus defining the performance limits of such kind of TI architecture. The component mismatches are generally characterized in the form of offset, gain, timing, and sampling bandwidth mismatches. Due to the periodic nature of time-interleaved architectures, they all appear in the form of undesired distortion tones around the frequency location of f_s/M in the signal spectrum and thus impose a limit in the achievable dynamic range of the system. For the offset and gain mismatches, self-calibration techniques [17–19] can be applied to correct the mismatch errors. However, the calibration of timing and bandwidth mismatches in time-interleaved systems is not an easy task since the errors that appeared are dynamic which depends on the input signal frequency in nature. Periodic timing skew can be caused by mismatched propagation delays in the multi-phase clock generator which provides time-interleaved sampling clock waveforms for the system, or due to undesired parasitic coupling (such as di/dt noise through the supply, generated by the switching of digital circuits [4]). While bandwidth mismatch would be generated by the mismatched in the timing-constant of the RC-sampling branch.

4.3   Channel Mismatch Analysis

All types of channel mismatches appeared as a form of errors that are periodic with frequency of $f_s/M = 1/(MT_s)$, e.g. the input signal being processed by the TI-ADC always pass through a specific channel at an interval of MT_s, i.e. it always see the same error in that channel for every M samples. As a result, channel mismatches in TI-ADC always create fixed tones (in the case of offset mismatches) or modulation sidebands (in gain, timing and bandwidth mismatches) at the locations which are multiples of f_s/M in the output spectrum of the ADC.

To analyze the channel mismatch effects, the mathematical representation of the output spectrum of the TI-ADC must be determined. Consider the TI-ADC samples a continuous-time signal $x(t)$ at nominally spaced time intervals of T_s, and then pass through the transfer function of the ADC to obtain the sampled sequence $y(t)$. Unlike other types of mismatches, timing-mismatch draws much more cares in TI sampled-data systems since different TI systems may have different timing characteristic. According to the classification in [5, 7], TI-ADC can be classified as **IN-OU(IS)** system, which means the Input signal is sampled by the system with Nonuniform time-interval and later played out or represented by discrete samples in the **O**utput at **U**niform time instants for later processing. This appeared typically in the analog to digital conversion path (timing-mismatch only @ input signal sampling), e.g. TI-ADCs [1] or multirate sampled-data decimators [2]. Since the output of this kind of systems is in the digital domain, the output is equivalently in Impulse-Sampled form and thus it is referred as **IN-OU(IS)** process for practical analog to digital conversion systems.

For a TI-ADC which is an IN-OU(IS) system, the time-domain output signal can be represented as:

$$y(t) = \sum_{n=-\infty}^{\infty} \{[x_n(t_n) * h_n(t_n)]\delta(t - nT)\} \tag{4.1}$$

where $h_n(t)$ is the system's impulse response and

$$x_n(t_n) = x(t_n) + O_n \tag{4.2}$$

which embeds the input referred offset O_n into the input signal $x(t_n)$ to simplify the analysis, and $*$ denotes the convolution operation. The subscript n means that the system impulse responses and input referred offsets are both vary sample-by-sample with period of M as a result of time-interleaved operation. For a TI system with M channels the following holds:

$$t_n = nT + \Delta t_n \tag{4.3}$$

where Δt_n is used to model the amount of timing mismatch which is a periodic sequence with period M, while the gain and bandwidth mismatch errors are model in the expression of $h_m(t)$ as presented in the subsequent subsections. To model the periodicity on Δt_n, let $n = kM + m$, which implies

$$t_n = (kM + m)T + \Delta t_{kM+m}$$

$$= kMT + mT + \Delta t_m$$

$$= kMT + mT + r_m T \tag{4.4}$$

where $r_m = \Delta t_m / T$ is the normalized periodic timing-skew sequence with period M, measured in percentage of T. Substituting (4.4) into (4.1) and utilizing the periodicity in t_n we have

$$y(t) = \sum_{m=0}^{M-1} \sum_{k=-\infty}^{\infty} \{(x_m * h_m)(kMT + mT + r_m T)\}\delta(t - kMT - mT) \tag{4.5}$$

Taking the Fourier Transform of (4.5) yields

$$Y(\omega) = \sum_{m=0}^{M-1} \left\{ \sum_{k=-\infty}^{\infty} \{[(x_m * h_m)(kMT + mT + r_m T)] \cdot e^{-j\omega kMT}\} e^{-j\omega mT} \right\}$$

$$= \sum_{m=0}^{M-1} \left\{ \sum_{k=-\infty}^{\infty} \frac{1}{2\pi} \left\{ \left[\int_{-\infty}^{\infty} X_m(j\Omega)H_m(j\Omega)e^{j\Omega mT}e^{j\Omega kMT}e^{j\Omega r_m T}d\Omega \right] e^{-j\omega kMT} \right\} e^{-j\omega mT} \right\}$$

$$= \frac{1}{2\pi} \sum_{m=0}^{M-1} \left\{ \left\{ \int_{-\infty}^{\infty} X_m(j\Omega) H_m(j\Omega) e^{j\Omega mT} e^{j\Omega r_m T} \cdot \left[\sum_{k=-\infty}^{\infty} e^{jk(\Omega-\omega)MT} \right] d\Omega \right\} e^{-j\omega mT} \right\}$$

$$(4.6)$$

and by applying the following identities

$$\sum_{k=-\infty}^{\infty} e^{jk(\Omega-\omega)MT} = \left(\frac{2\pi}{MT} \right) \sum_{k=-\infty}^{\infty} \delta\left(\Omega - \omega + k\frac{2\pi}{MT} \right) \qquad (4.7)$$

The output spectrum can be simplified to

$$Y(\omega) = \frac{1}{MT} \sum_{m=0}^{M-1}$$

$$\left\{ \left\{ \sum_{k=-\infty}^{\infty} \int_{-\infty}^{\infty} X_m(j\Omega) H_m(j\Omega) e^{j\Omega mT} e^{j\Omega r_m T} \delta\left(\Omega - \omega + k\frac{2\pi}{MT}\right) d\Omega \right\} e^{-j\omega mT} \right\}$$

$$= \frac{1}{MT} \sum_{m=0}^{M-1} \left\{ \left\{ \sum_{k=-\infty}^{\infty} X_m\left(\omega - k\frac{2\pi}{MT}\right) H_m\left(\omega - k\frac{2\pi}{MT}\right) e^{j(\omega - k\frac{2\pi}{MT})(mT + r_m T)} \right\} e^{-j\omega mT} \right\}$$

$$= \frac{1}{T} \sum_{k=-\infty}^{\infty} A_k(\omega) X_m\left(\omega - k\frac{2\pi}{MT} \right)$$

$$= \frac{1}{T} \sum_{k=-\infty}^{\infty} \left[A_k(\omega) X\left(\omega - k\frac{2\pi}{MT} \right) + B_k \delta\left(\omega - k\frac{2\pi}{MT} \right) \right] \qquad (4.8)$$

where

$$A_k(\omega) = \frac{1}{M} \sum_{m=0}^{M-1} H_m\left(\omega - k\frac{2\pi}{MT} \right) e^{j\omega r_m T} e^{-jk r_m \frac{2\pi}{M}} e^{-jkm\frac{2\pi}{M}} \qquad (4.9)$$

represents the weights of the modulation sidebands of input signal spectrum $X(\omega)$ and

$$B_k = \frac{1}{M} \sum_{m=0}^{M-1} O_m H_m(0) e^{-jkm\frac{2\pi}{M}} \qquad (4.10)$$

represents the weights of fixed offset induced tones located at multiples of $1/MT = f_s/M$. Equations (4.8)–(4.10) completely describe the spectrum of a TI-ADC system including all the mismatches. It shows that the characteristic of all types of mismatches are similar, also creating weighted modulation sidebands or offset tones at multiples of f_s/M (with the weights specified by A_k and B_k in (4.9)), as shown in Fig. 4.2 for an example of $M = 4$. Notice that A_k and B_k are both periodic with period M, and $k = 0$ in (4.8), (4.9) corresponding to the intended output spectrum, since

$$A_0(\omega) = \frac{1}{M} \sum_{m=0}^{M-1} H_m(\omega) e^{j\omega r_m T} \approx \frac{1}{M} \sum_{m=0}^{M-1} H_m(\omega) \qquad (4.11)$$

for small value of r_m (which is practical as the amount of timing-skew should be small compared to the whole sampling period T), i.e. the original signal is now shaped by the average of the transfer functions in different channels. The overall channel offset term in (4.10) for $k = 0$ can be simplified as

$$B_0 = \frac{1}{M} \sum_{m=0}^{M-1} O_m H_m(0) \qquad (4.12)$$

that means the overall channel offset is also the average of the channel offsets (which equals to input referred offset shaped by individual channel transfer function at DC).

Although (4.9) represents the complete spectrum description of a TI-ADC with channel mismatches, it presents less information about the amount of the mismatches that can be tolerated under specific system performance which is mandatory for the early design phase. Specifically, it is more informative to evaluate the SNDR of such system due to the various types of mismatches independently, which will be presented in subsequent sections.

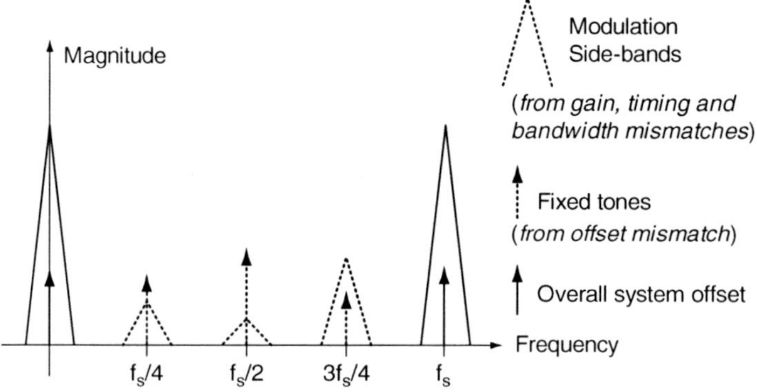

Fig. 4.2 General output spectrum of time-interleaved ADC ($M = 4$)

4.4   Offset Mismatch

Offset mismatches among TI channels are generated by the input referred offsets of the whole sub-ADCs, which are statistically distributed among different channels. The offset may be contributed from the input referred offset from the differential pair of the opamps, which can be compensated through various types of auto-zeroing technique (like the low-voltage offset compensation scheme presented in Chapter 3). In additions, offset can also be generated by capacitor mismatches if the capacitors are used to process common-mode level shifts (like the voltage-control level-shifting techniques presented in Chapter 3).

To evaluate the effects of offset mismatches to the performance of the TI-ADC, it is constructive to neutralize other types of channel mismatch errors. This is equivalent to set $h_m(t) = h(t)$ for all m to disable gain and bandwidth mismatch errors, and $r_m = 0$ for all m to immobilize the timing mismatch errors in (4.8) to (4.10). Equation (4.9) then can be simplified as:

$$A_k(\omega) = \frac{1}{M} H\left(\omega - k\frac{2\pi}{MT}\right) \sum_{m=0}^{M-1} e^{-jkm\frac{2\pi}{M}}$$

$$= \begin{cases} H(\omega - k\frac{2\pi}{MT}), & \text{for } k = 0, M, 2M \ldots \\ 0 & , \text{ for } k \neq 0, M, 2M \ldots \end{cases} \qquad (4.13)$$

and Eq. (4.8) can be rearranged as:

$$Y(\omega) = \frac{1}{T} \sum_{k=-\infty}^{\infty} \left[X\left(\omega - k\frac{2\pi}{T}\right) H\left(\omega - k\frac{2\pi}{T}\right) + B_k\delta\left(\omega - k\frac{2\pi}{MT}\right) \right] \qquad (4.14)$$

where

$$B_k = \frac{1}{M} H(0) \sum_{m=0}^{M-1} O_m e^{-jkm\frac{2\pi}{M}} \qquad (4.15)$$

Equation (4.14) shows that the output spectrum contains the intended discrete-time signal spectrum plus the fixed offset tones with weights defined in (4.15) which are not signal-dependent.

Assuming that the channel offsets O_m are independent, identically distributed Gaussian random variables with zero mean and a standard deviation of σ_{os}, then the expected power of offset tones can be evaluated as follows:

$$E\left[|B_k|^2\right] = E\left[B_k B_k^*\right] = \frac{1}{M^2} |H(0)|^2 \sum_{m=0}^{M-1} \sum_{n=0}^{M-1} E[O_m O_n] e^{-jk(m-n)\frac{2\pi}{M}} \qquad (4.16)$$

Notice that $E[O_m O_n]$ always evaluates to zero unless $m = n$ since O_m and O_n are two uncorrelated, zero-mean random variables, and (4.16) can be rewritten as

$$E\left[|B_k|^2\right] = \frac{|H(0)|^2}{M^2} \sum_{m=0}^{M-1} E\left[O_m^2\right] = \frac{|H(0)|^2}{M^2} \sum_{m=0}^{M-1} \sigma_{os}^2 = \frac{|H(0)|^2}{M} \sigma_{os}^2 \qquad (4.17)$$

Equation (4.17) shows that the all the offset tones have identical expected or mean power. Without counting the overall system offset at $k = 0$ at DC which induced no distortion to the system, the distortion power due to offset mismatch can be evaluated as

$$P_{os} = \sum_{k=1}^{M-1} E\left[|B_k|^2\right] = (M - 1)\frac{|H(0)|^2}{M} \sigma_{os}^2 = \left(1 - \frac{1}{M}\right)|H(0)|^2 \sigma_{os}^2 \qquad (4.18)$$

Consider a sinusoidal input signal with frequency of $\omega_0 = 2\pi f_0$ and amplitude of A, with the average sampled signal power of

$$P_{signal} = \frac{A^2}{2}|H(\omega_0)|^2 \qquad (4.19)$$

The SNDR due to offset mismatch can be calculated as

$$SNDR_{offset\_mis} = 10 \log_{10}\left(\frac{P_{signal}}{P_{os}}\right)$$
$$= 20 \log\left(\frac{A}{\sqrt{2}\sigma_{os}} \cdot \left|\frac{H(\omega_0)}{H(0)}\right|\right) - 10 \log_{10}\left(1 - \frac{1}{M}\right) \qquad (4.20)$$

which can be used to evaluate the allowed standard deviation of input-referred offset in each sub-ADC. Notice that if we treat the offset located at DC (i.e. $k = 0$) as distortion also, then there are total M offset tones instead of M-1 tones and (4.20) reduced to

$$SNDR_{offset\_mis\_v2} = 20 \log\left(\frac{A}{\sqrt{2}\sigma_{os}} \cdot \left|\frac{H(\omega_0)}{H(0)}\right|\right) \qquad (4.21)$$

which is identical to the traditional SNDR formula in [8] with $H(\omega) = 1$ for all ω. Figure 4.3 shows a plot of (4.20) and (4.21) versus no. of TI channels M to evaluate the difference between two equations ($A = 1$ V, $\sigma_{os} = 1$ mV and $H(\omega) = 1$ for all ω). The SNDR difference can be as large as 3 dB for two-channel TI ADC since there are only two offset tones, one at DC and one at Nyquist. For large number of M, (4.20) converges to (4.21). For general TI-ADC (4.20) can be used since an overall DC offset can be easily removed by the system, however the offset mismatch cannot.

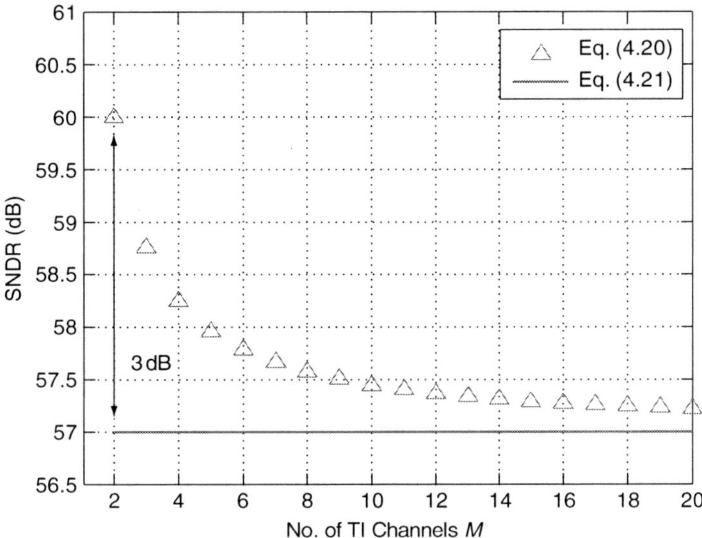

Fig. 4.3 Comparison of SNDR for offset mismatch using (4.20) and (4.21)

As a final remark, (4.20) also includes the channel transfer function in the equation, and it is clearly seen that although the offset mismatch tones appeared at frequencies which are multiples of f_s/M, they are all shaped by the sub-ADC's channel transfer function at DC frequency, that means they cannot be filtered by the individual channel transfer function, e.g. a low-pass sampling network in the sub-ADC cannot filter-out the offset mismatch tone at Nyquist. This can be explained as individual channel offset appeared as DC signal inside the individual channel.

4.5 Gain Mismatch

Gain mismatches in time-interleaved systems are originated by the mismatch in channel's DC-gain, which may be contributed from the capacitor mismatches as well as finite-opamp gain error among channels. Notice that dynamic-type (frequency-dependent) gain-mismatch are more difficult to characterized since the performance of such system will heavily depends on the frequency response of the channel, but the mismatch in first order low-pass transfer function (known as bandwidth-mismatch) can still be analyzed which typically existed in the RC sampling front-end of the TI-ADC. The effect of static gain mismatch will be considered first and then followed by the bandwidth mismatch in later section.

The effects of gain mismatch can be evaluated by setting $H_m(\omega) = H(\omega)(1 + \delta_m)$ to enable gain but disable bandwidth mismatch errors, and $O_m = r_m = 0$ for all m

to immobilize the offset and timing mismatch errors in (4.8) to (4.10). As a result $B_k = 0$ for all k in (4.10), Eqs. (4.8) and (4.9) can be simplified as:

$$Y(\omega) = \frac{1}{T} \sum_{k=-\infty}^{\infty} A_k(\omega)X\left(\omega - k\frac{2\pi}{MT}\right) \tag{4.22}$$

where

$$A_k(\omega) = \frac{1}{M}H\left(\omega - k\frac{2\pi}{MT}\right) \sum_{m=0}^{M-1} (1 + \delta_m)e^{-jkm\frac{2\pi}{M}} \tag{4.23}$$

Thus the spectrum of the TI-ADC under gain mismatch consisted of weighted (by A_k) modulation sidebands of original input spectrum at frequency multiples of f_s/M. However as shown in (4.23), the weights are not input-frequency-dependent, thus the performance of gain mismatch does not degrade as the signal frequency increases.

The SNDR of the TI-ADC under gain mismatch can be evaluated from the weights in (4.23). For an input sinusoidal signal with frequency $\omega_0 = 2\pi f_0$, the sidebands contain tones located at $\omega = \pm\omega_0 + k(2\pi)/(MT)$, with $k = 0$ representing the original input signal and $k = +1, +2, \ldots$ or $k = -1, -2, \ldots$ the positive and negative image components, respectively. Due to the symmetry property of the Fourier Transform of real signals, the images at positive frequencies of $\omega = -\omega_0 + k(2\pi)/(MT)$ with $k = +1, +2, \ldots$ will be directly reflected by the images at negative frequencies of $\omega = \omega_0 + k(2\pi)/(MT)$ with $k = -1, -2, \ldots$. Thus, the calculation can be simplified by evaluating the sideband components at only $\omega = \omega_0 + k(2\pi)/(MT)$ over the range $[-f_s/2, f_s/2]$ (or equivalently $[0, f_s]$) to find the SNDR (with respect to the signal only centered at ω_0) over the range of $[0, f_s/2]$. In the following formula's derivation it has been assumed that $\omega_0 \neq k(2\pi)/(MT)$, implying that the signal (and the sidebands as well) is (are) not exactly located at an integer multiple of f_s/M. Notice that all these assumptions are also valid for subsequent timing and bandwidth mismatch analysis.

Simplifying (4.23) using $\omega = \omega_0 + k(2\pi)/(MT)$ it yields:

$$A_k\left(\omega_0 + k\frac{2\pi}{MT}\right) = \frac{1}{M}H(\omega_0) \sum_{m=0}^{M-1} (1 + \delta_m)e^{-jkm\frac{2\pi}{M}} \tag{4.24}$$

where for $k = 0$ $A_0(\omega_0)$ corresponds to the signal component, and for $k = 1$ to M-1 A_k correspond to M-1 different sideband components. The SNDR, over the range of $[0, f_s]$ or equivalently for $k = 1$ to M-1, can be expressed as [20]:

$$SNDR = 10 \log_{10} \left[\frac{E[|A_0(\omega_o)|^2]}{E\left[\sum_{k=1}^{M-1} \left|A_k\left(\omega_o + k\frac{2\pi}{MT}\right)\right|^2\right]} \right] dB \tag{4.25}$$

Assuming that the channel gain-mismatch components δ_m are independent, identically distributed Gaussian random variables with zero mean and a standard deviation of σ_δ, then the expected power of the sidebands can be evaluated as follows:

$$E\left[\left|A_k\left(\omega_0 + k\frac{2\pi}{MT}\right)\right|^2\right] = \frac{|H(\omega_0)|^2}{M^2} \sum_{m=0}^{M-1}\sum_{n=0}^{M-1} E[(1+\delta_m)(1+\delta_n)]e^{-jk(m-n)\frac{2\pi}{M}}$$

$$= \frac{|H(\omega_0)|^2}{M^2} \sum_{m=0}^{M-1}\sum_{n=0}^{M-1} \{E[1] + E[\delta_m\delta_n] + E[(\delta_m + \delta_n)]\}e^{-jk(m-n)\frac{2\pi}{M}}$$

$$= \frac{|H(\omega_0)|^2}{M^2} \left\{\sum_{m=0}^{M-1}\sum_{n=0}^{M-1} E[1]e^{-jk(m-n)\frac{2\pi}{M}} + \sum_{m=0}^{M-1}\sum_{n=0}^{M-1} E[\delta_m\delta_n]e^{-jk(m-n)\frac{2\pi}{M}}\right\}$$

$$= \frac{|H(\omega_0)|^2}{M^2}\{0 + M\sigma_a^2\} = \frac{|H(\omega_0)|^2}{M}\sigma_\delta^2, \quad \text{for } k \neq 0,\ M,\ 2M\ldots \tag{4.26}$$

Similar to the case of offset mismatch, (4.26) also shows that the all the sidebands have identical expected or mean power. The total distortion power due to gain mismatch can be evaluated as

$$P_{dist} = \sum_{k=1}^{M-1} E\left[\left|A_k\left(\omega_0 + k\frac{2\pi}{MT}\right)\right|^2\right] = (M-1)\frac{|H(\omega_0)|^2}{M}\sigma_\delta^2$$

$$= \left(1 - \frac{1}{M}\right)|H(\omega_0)|^2\sigma_\delta^2 \tag{4.27}$$

The expected power of the signal component can be calculated as:

$$E\left[|A_0(\omega_0)|^2\right] = \frac{|H(\omega_0)|^2}{M^2} \sum_{m=0}^{M-1}\sum_{n=0}^{M-1} E[(1+\delta_m)(1+\delta_n)]$$

$$= \frac{|H(\omega_0)|^2}{M^2} \left\{\sum_{m=0}^{M-1}\sum_{n=0}^{M-1} E[1] + \sum_{m=0}^{M-1}\sum_{n=0}^{M-1} E[\delta_m\delta_n]\right\}$$

$$= \frac{|H(\omega_0)|^2}{M^2}\{M^2 + M\sigma_\delta^2\} \approx |H(\omega_0)|^2 \tag{4.28}$$

The last step of (4.28) hold since the standard deviation of gain mismatch σ_δ is small compared with the nominal gain. The SNDR due to gain mismatch can be calculated by substituting (4.26) and (4.28) into (4.25) as:

$$SNDR_{gain\_mis} = 20 \log\left(\frac{1}{\sigma_\delta}\right) - 10 \log_{10}\left(1 - \frac{1}{M}\right) \tag{4.29}$$

which corresponds to the results in [8] (with the mapping of $\sigma_\delta = \sigma_a/a$ in [8]), however the results derived in here shows that the performance of the gain mismatch does not depends on the specific channel transfer characteristic $H(\omega)$ as long as the gain mismatch are of static type. Actually as shown in (4.26) all the sidebands are shaped by the transfer function at input signal frequency, although the sidebands themselves are located at frequency multiples of f_S/M. The reason is that the individual channel in fact only processes the original input signal frequency, so the mismatch tones are also shaped by $H(\omega_0)$.

4.6 Timing Mismatch

Timing mismatch (or known as periodic timing-skew) in time-interleaved systems refers to the mismatch in sampling instant inside the individual channel. They are originated from the inaccurate sampling clock edges generated by the mismatch in the clock generation paths. Timing mismatches are difficult to compensate due to its signal-dependent dynamic nature, and complex algorithms [21, 22] are generally required to calibrate the timing mismatch in digital domain which leads to large digital area and power consumption.

The effects of timing mismatch can be evaluated by setting $H_m(\omega) = H(\omega)$ and $O_m = 0$ for all m to deactivate all the other type of mismatch errors in (4.8) to (4.10). Then $B_k = 0$ for all k in (4.10), and (4.9) can be simplified as:

$$A_k(\omega) = \frac{1}{M} H\left(\omega - k\frac{2\pi}{MT}\right) \sum_{m=0}^{M-1} e^{j\omega r_m T} e^{-jkr_m \frac{2\pi}{M}} e^{-jkm\frac{2\pi}{M}} \tag{4.30}$$

Again the spectrum under timing mismatch consisted of weighted (by A_k) modulation sidebands of original input spectrum at frequency multiples of f_s/M. However, unlike the offset or gain mismatch, the weights are now input-frequency-dependent, thus the effect of timing mismatch are more pronounced at high input-signal frequency (but not the sampling frequency).

To evaluate the SNDR, assume a sinusoidal input signal and simplifying (4.30) using $\omega = \omega_0 + k(2\pi)/(MT)$ it yields:

$$A_k\left(\omega_0 + k\frac{2\pi}{MT}\right) = \frac{1}{M} H(\omega_0) \sum_{m=0}^{M-1} e^{j\omega_0 r_m T} e^{-jkm\frac{2\pi}{M}} \tag{4.31}$$

Assuming that the timing-mismatch components r_m are independent, identically distributed Gaussian random variables with zero mean and a standard deviation of σ_{r_m}.

To evaluate the expected value of sidebands power the following statistical formula can be utilized [23]:

$$E[e^{j\omega_0 T r_m}] = e^{-\sigma_{rm}^2 \omega_o^2 T^2/2} \tag{4.32}$$

Then the expected power of the sidebands can be evaluated as follows:

$$E\left[\left|A_k\left(\omega_0 + k\frac{2\pi}{MT}\right)\right|^2\right] = \frac{|H(\omega_0)|^2}{M^2} \sum_{m=0}^{M-1}\sum_{n=0}^{M-1} E\left[e^{j\omega_0(r_m - r_n)T}\right] e^{-jk(m-n)\frac{2\pi}{M}}$$

$$= \frac{|H(\omega_0)|^2}{M^2}\left\{M + (0-M)e^{-\omega_0^2\sigma_{rm}^2 T^2}\right\} \approx \frac{|H(\omega_0)|^2}{M}\omega_0^2\sigma_{rm}^2 T^2, \text{ for } k \neq 0, M, 2M\ldots \tag{4.33}$$

for small value of r_m and thus σ_{rm}. Similarly all the sidebands have identical expected or mean power. The total distortion power due to timing mismatch can be evaluated as

$$P_{dist} = \sum_{k=1}^{M-1} E\left[\left|A_k\left(\omega_0 + k\frac{2\pi}{MT}\right)\right|^2\right] = (M-1)\frac{|H(\omega_0)|^2}{M}\omega_0^2\sigma_{rm}^2 T^2$$

$$= \left(1 - \frac{1}{M}\right)|H(\omega_0)|^2\omega_0^2\sigma_{rm}^2 T^2 \tag{4.34}$$

The expected power of the signal component can be calculated as:

$$E\left[|A_0(\omega_0)|^2\right] = \frac{|H(\omega_0)|^2}{M^2} \sum_{m=0}^{M-1}\sum_{n=0}^{M-1} E\left[e^{j\omega_0(r_m - r_n)T}\right]$$

$$= \frac{|H(\omega_0)|^2}{M^2}\left\{M + (M^2 - M)e^{-\omega_0^2\sigma_{rm}^2 T^2}\right\}$$

$$\approx \frac{|H(\omega_0)|^2}{M^2}\left\{M^2 - (M^2 - M)\omega_0^2\sigma_{rm}^2 T^2\right\} \approx |H(\omega_0)|^2 \tag{4.35}$$

The SNDR due to timing mismatch can be calculated by substituting (4.34) and (4.35) into (4.25) as:

$$SNDR_{timing\_mis} = 20\ \log\left(\frac{1}{2\pi f_0 \sigma_{rm} T}\right) - 10\ \log_{10}\left(1 - \frac{1}{M}\right)$$

$$= 20 \log\left(\frac{1}{2\pi f_0 \sigma_t}\right) - 10 \log_{10}\left(1 - \frac{1}{M}\right) \tag{4.36}$$

where $\sigma_t = \sigma_{rm} T$ represents the standard deviation of the timing mismatch in unit of second. From (4.36) it is evidence that the performance of the TI-ADC under timing mismatch degrades as the input signal frequency increases, but it has no relation to the sampling frequency since usually the amount of timing mismatch is specified as absolute value of second. Equation (4.36) corresponds to the results in [8, 10], however the derivation shown in here presented important information that the performance again does not depend on the specific characteristic of the channel transfer function.

4.7 Bandwidth Mismatch

Bandwidth mismatches are originated from the mismatch in sampling timing-constant in the front-end of TI sampling systems. A typical S/H circuit widely used in the front-end of sampled-data systems is shown in Fig. 4.4, which must also be part of each channel of the ADC front-end of Fig. 4.1. In phase 1 the continuous-time input signal is passively sampled into the capacitor C and also RC-filtered by the finite bandwidth imposed by the sampling switch's on-resistance R and the sampling capacitor C. The mismatches occurring in the sampling switches and the capacitors among various channels will lead to the overall bandwidth mismatch. Notice that this kind of bandwidth mismatch will only modulate the continuous-time input signal, and after the signal being sampled into the capacitor C the operation of the remaining part of the S/H is in discrete-time and all the mismatches will appear as offset and pure gain-mismatches that were already fully character-ized before.

A thorough spectra-domain analysis of the bandwidth mismatch effect will be proposed here and based on this a closed-form SNDR expression will be derived [6]. In traditional sense, one way to avoid the limitation from bandwidth mismatch among the channels is to increase the sampling bandwidths, so that they are greater than the maximum input frequency to be digitized. However, it will be proven that the most important contribution for the bandwidth mismatch is the phase mismatch originated from the phase-shift of the RC-filtering by the sampling branch, which

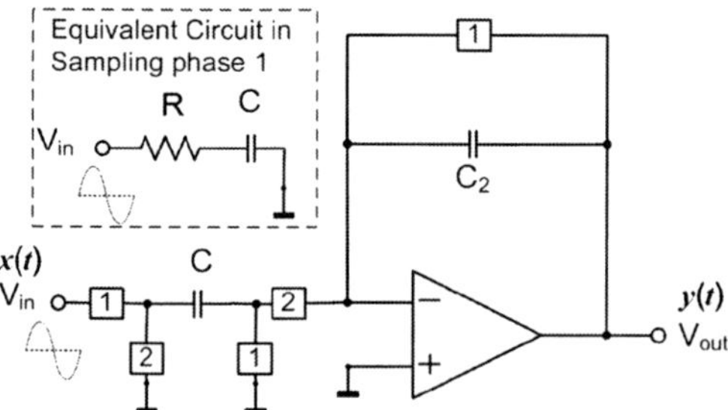

Fig. 4.4 A typical front-end S/H circuit

causes performance degradation even for signal frequencies smaller than the bandwidth of such sampling branch, thus weaken the effectiveness of the increased sampling bandwidth (a quite unexpected result). Although the analysis of deterministic bandwidth mismatches had already been addressed in [9, 12] and [13], this section will present a comprehensive and thorough study with a statistical approach that comprises the following improved features over previous works:

(a) A handy closed-form SNDR expression is derived.
(b) The formula can be generalized for any number (M) of TI channels.
(c) Since intrinsic circuit imbalance (being highly circuit- and architecture-dependent) or inadequate layout design will introduce systematic mismatches, and considering that only standard deviations of various mismatch parameters are given by the foundries process datasheets, a statistical method dealing with random mismatches is derived.

The formula derived here will simplify the performance evaluation of TI sampling systems under bandwidth mismatches, and will also allow the prior determination of the specifications of sampling switches and capacitors in the design phase for a specifically targeted value of SNDR. Comparison with the analysis in [14] which focuses on the design methodology for digital post calibration that valid for small input signal bandwidth, the analysis here is targeting a prior analysis to suppress the bandwidth mismatch within a full input signal bandwidth at a much earlier design phase that is well suited for usual top-down analog design flows.

Assuming that R_m and C_m are the switch's on-resistance and the sampling capacitance of the mth channel, respectively, then, the sampling time constant would be $\tau_m = R_m C_m = \tau(1 + \tau_m)$ where τ corresponds to its mean/nominal value and τ_m is the percentage of the time-constant mismatch, which are Gaussian distributed random variables with zero mean and standard deviation of $\sigma_{\tau m}$. Based on this, the frequency response in the mth channel can be defined as:

$$H_m(\omega) = \frac{a}{1 + j\omega\tau_m} = \frac{a}{1 + j\omega\tau(1 + \tau_m)} \tag{4.37}$$

where a is the dc gain which is identical for all channels, so the gain mismatch error is neutralized and only bandwidth mismatch is considered (Also $O_m = 0$ and $r_m = 0$). Using $\omega = \omega_0 + k(2\pi)/(MT)$ the weights of modulation sidebands in (4.9) can be simplified as:

$$A_k\left(\omega_0 + k\frac{2\pi}{MT}\right) = \frac{1}{M}\sum_{m=0}^{M-1} H_m(\omega_0)e^{-jkm\frac{2\pi}{M}} = \frac{a}{M}\sum_{m=0}^{M-1}\frac{e^{-jkm\frac{2\pi}{M}}}{1 + j\omega\tau(1 + \tau_m)} \tag{4.38}$$

Following the formula derivations in Appendix B the SNDR for the TI-ADC with bandwidth mismatch can be proven as:

$$SNDR_{banwidth\_mis} = 10\,\log_{10}\left[1 + \omega_0^2\tau^2\right] - 20\,\log_{10}\left[\omega_0\tau\sigma_{\tau m}\right]$$

$$- 10\,\log_{10}\left[1 - \frac{1}{M}\right]dB \qquad\qquad (4.39)$$

Equation (4.39) can be used for quick evaluation of the performance of TI sampled-data systems under the influence of bandwidth mismatches. Important information can be extracted directly from (4.39), namely that SNDR is inversely proportional to the signal frequency, as expected. Notice that the SNDR only degrades by an amount of 3 dB from $M = 2$ to ∞, as shown in Fig. 4.5, a fact that was not explained before when the phenomenon of bandwidth mismatch was previously analyzed [9, 12] and [13] (actually a similar conclusion can be found in the offset, gain and timing mismatch analysis).

To verify the accuracy of the formula MATLAB FFT model simulations are presented. Figure 4.6a shows plots of the simulated SNDR versus normalized signal frequency $\omega_0\tau$ and $\sigma_{\tau m}$ by 10,000 times Monte-Carlo simulations under $M = 4$ and $\omega_0 T = 2\pi \times 0.2495$, with the absolute error (in dB) between the simulated and calculated SNDR presented in Fig. 4.6b. From this figure the absolute error between the simulated and calculated SNDR is well below 0.5 dB for the normal range of $\omega_0\tau$ and $\sigma_{\tau m}$. The error starts to increase with higher values of the mismatch $\sigma_{\tau m}$ and high input frequency, since an approximation of $\sigma_{\tau m}\!\ll\!1$ is considered in the previous formula derivations, but such a large value of mismatches (e.g. >10%) are not typical in CMOS which will lead to an SNDR of only <20dB. Figure 4.7a shows another 3D plots of the simulated SNDR versus the number of TI channels M and $\sigma_{\tau m}$ by 10,000 times Monte-Carlo simulations under $\omega_o\tau = 0.8$ and

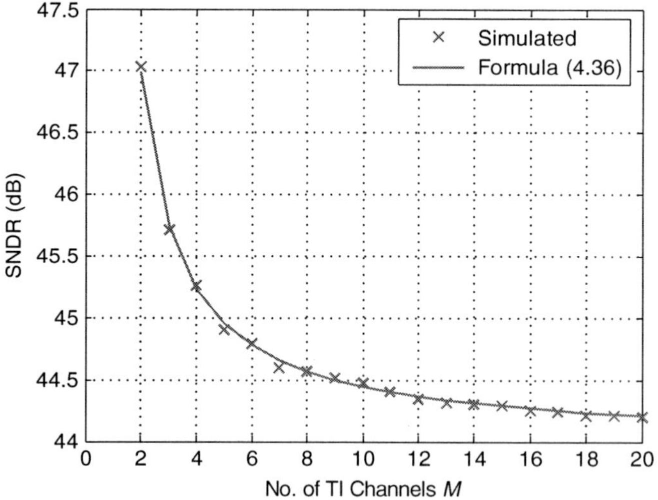

Fig. 4.5 A plot of no. of TI channels M versus SNDR (simulated and calculated from (4.36) with $\omega_0\tau = 0.8$ and $\sigma_{\tau_m} = 1\%$

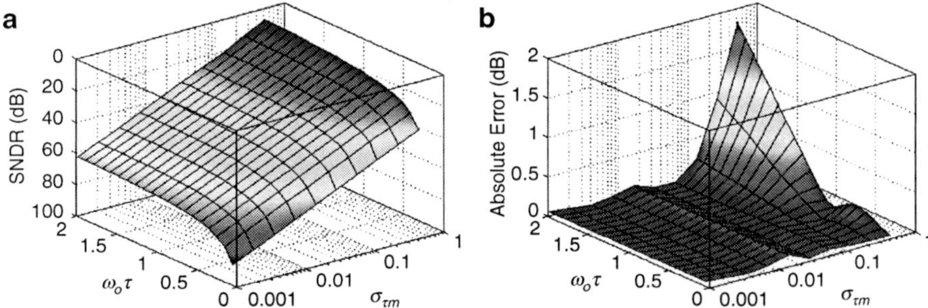

Fig. 4.6 (**a**) Simulated SNDR and (**b**) absolute error between the simulated and calculated SNDR of TI systems under bandwidth mismatches versus normalized frequency $\omega_0\tau$ and mismatch standard derivation σ_{τ_m} by 10,000 times Monte Carlo Simulations ($M = 4$, $\omega_0 T = 2\pi \times 0.2495$)

$\omega_0 T = 2\pi \times 0.2495$, with the absolute error (in dB) between the simulated and calculated SNDR presented in Fig. 4.7b. It is also demonstrated that the formula can accurately predict systems' performance under any number of TI channels M.

The first order filtering effect caused by the sampling branch will usually have more influence when the signal frequency is near or greater than its bandwidth, since before the -3 dB-frequency the filter provides only negligible attenuation. However, the bandwidth mismatches among TI-channels can have a large impact on the performance even when the signal frequency is well below the corner-frequency (quite unexpectedly), since the phase shift starts to increase at 1/10 of the corner-frequency. To demonstrate this effect, consider an idealized RC-filter which do not have phase response but only have the magnitude response:

$$H_{m\_magnitude\_only}(\omega) = \frac{1}{\sqrt{1 + (\omega\tau_m)^2}} = \frac{1}{\sqrt{1 + [\omega\tau(1 + \tau_m)]^2}} \tag{4.40}$$

Then the corresponding SNDR can be calculated as

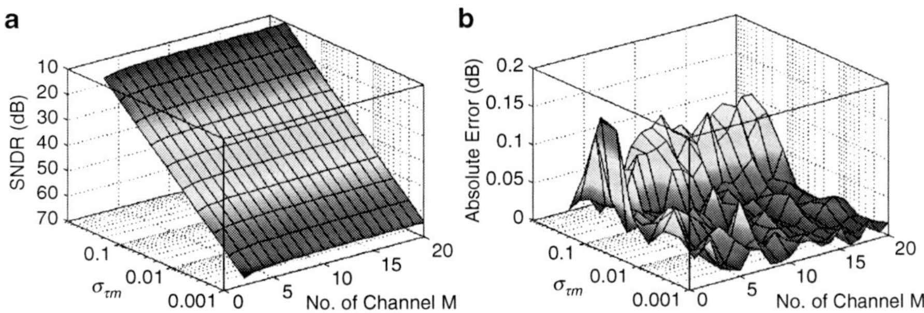

Fig. 4.7 (**a**) Simulated SNDR and (**b**) absolute error between the simulated and calculated SNDR of TI systems under bandwidth mismatches versus no. of TI channels M and mismatch standard derivation σ_{τ_m} by 10,000 times Monte Carlo Simulations ($\omega_0\tau = 0.8$, $\omega_0 T = 2\pi \times 0.2495$)

$$SNDR_{magnitude\_only} = 10 \log_{10}\left[1 + \omega_0^2\tau^2\right] - 20 \log_{10}\left[\omega_0\tau\sigma_{\tau m}\right]$$

$$-10 \log_{10}\left[1 - \frac{1}{M}\right] - 10 \log_{10}\left[\frac{\omega_0^2\tau^2}{1 + \omega_0^2\tau^2}\right] dB \quad (4.41)$$

which differs from (4.39) only in the last term. Figure 4.8 presents a plot of (4.39) and (4.41) as a function of $\omega_0\tau$ with $M = 4$ and $\sigma_{\tau m} = 1\%$. It clearly shows that the effect of the phase mismatch is dominant especially at low frequency, which will lead to (e.g. 25 dB) overestimation of SNDR for the TI systems if only gain mismatch is considered as a dominating factor below the corner frequency.

4.8 Summary

In this chapter the principles and practical considerations of the design of time-interleaved ADCs have been investigated. Various types of mismatches (including offset, gain, timing and bandwidth mismatches) among different time-interleaved channels will create fixed pattern tones or modulation sidebands which will degrade the performance of such ADC systems.

The effects of all types of mismatches in the system performance had been thoroughly investigated in the spectral domain, and closed form expressions for the SNDR of the TI-ADC are derived as follows:

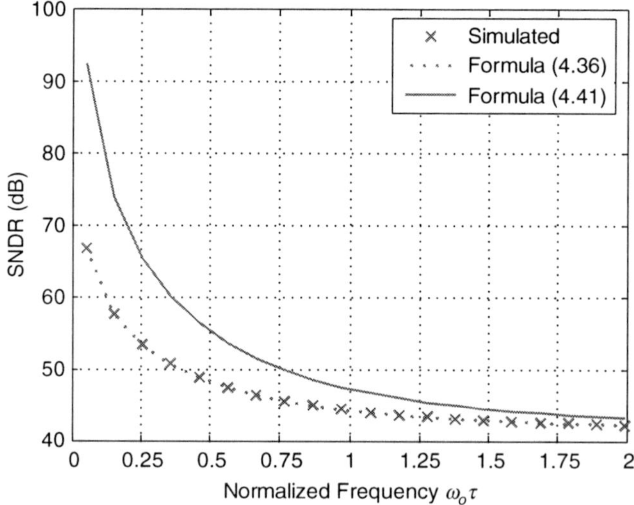

Fig. 4.8 A plot of normalized frequency $\omega_0\tau$ versus SNDR (simulated and calculated from (4.36) and (4.41) with $M = 4$ and $\sigma_{\tau_m} = 1\%$

$$SNDR_{offset\_mis} = 10\ \log_{10}\left(\frac{P_{signal}}{P_{os}}\right)$$

$$= 20\ \log\left(\frac{A}{\sqrt{2}\sigma_{os}} \cdot \left|\frac{H(\omega_0)}{H(0)}\right|\right) - 10\ \log_{10}\left(1 - \frac{1}{M}\right)$$

$$SNDR_{gain\_mis} = 20\ \log\left(\frac{1}{\sigma_\delta}\right) - 10\ \log_{10}\left(1 - \frac{1}{M}\right)$$

$$SNDR_{timing\_mis} = 20\ \log\left(\frac{1}{2\pi f_0 \sigma_t}\right) - 10\ \log_{10}\left(1 - \frac{1}{M}\right)$$

$$SNDR_{banwidth\_mis} = 10\ \log_{10}\left[1 + \omega_0^2\tau^2\right] - 20\ \log_{10}[\omega_0\tau\sigma_{\tau m}] - 10\ \log_{10}\left[1 - \frac{1}{M}\right] dB$$

These formulas can be used to determine the amount of the various types of mismatches that can be tolerated during the early design phase, thus allowing to derive the corresponding specification in the design of the various circuit blocks, e.g. the offset specifications of the opamps, the matching of the capacitor ratio and thus the size of the capacitors, the amount of timing-skew that is allowed in the design of clock generators and clock buses, as well as the size of switches, sampling capacitors, and the matching requirements in the layout routing of the input signal and clock buses.

References

1. L. Sumanen, M. Waltari, K. Halonen, A 10-bit 200MS/s CMOS parallel pipeline A/D converter, in *Proceedings of ESSCIRC* (Sept 2000), pp. 440–443
2. R.F. Neves, J.E. Franca, A CMOS Switched-Capacitor bandpass filter with 100 MSample/s input sampling and frequency down-conversion, in *Proceedings of ESSCIRC* (Sept 2000), pp. 248–251
3. C.-K.K. Yang et al., A Serial-Link Transceiver Based on 8-GSamples/s A/D and D/A Converters in 0.25-μm CMOS. in *IEEE J. Solid-St. Circ.* **36**(11), 1684–1692 (Nov 2001)
4. U. Seng-Pan, R.P. Martins, J.E. Franca, A 2.5V, 57MHz, 15-Tap SC bandpass interpolating filter with 320MHz output sampling rate in 0.35μm CMOS, in *ISSCC Digest of Technical Papers* (Feb 2002), pp. 380–382
5. U. Seng-Pan, S.-W. Sin, R.P. Martins, Exact spectra analysis of sampled signal with jitter-induced nonuniformly holding effects. in *IEEE Trans. Instrum. Meas.* **53**(4), 1279–1288 (Aug 2004)
6. S.-W. Sin, U.-F. Chio, U. Seng-Pan, R.P. Martins, Statistical spectra and distortion analysis of time-interleaved sampling bandwidth mismatch. in *IEEE T. Circuits Syst. II* **55**(7), 648–652 (July 2008)
7. S.-W. Sin, *Timing-Jitter Effects in Time-Interleaved Sampled-Data Systems – Generalized Noise Analysis & Mismatch-Insensitive Clock Generation*, Master Thesis, University of Macau, Macao SAR, China, 2003

8. M. Gustavsson et al., *CMOS Data Converters for Communications* (Kluwer, Boston, MA, 2000)
9. N. Kurosawa, H. Kobayashi, K. Maruyama, H. Sugawara, K. Kobayashi, Explicit analysis of channel mismatch effects in time-interleaved ADC systems. in *IEEE T. Circuits Syst. I* **48**(3), 261–271 (March 2001)
10. Y.C. Jenq, Digital spectra of nonuniformly sampled signals: fundamentals and high-speed waveform digitizers. in *IEEE Trans. Instrum. Meas.* **37**(2), 245–251 (June 1988)
11. Y.C. Jenq, Digital-to-Analog (D/A) converters with nonuniformly sampled signals. in *IEEE Trans. Instrum. Meas.* **45**(1), 56–59 (Feb 1996)
12. T.-H. Tsai, P.J. Hurst, S.H. Lewis, Bandwidth mismatch and its correction in time-interleaved analog-to-digital converters. in *IEEE T. Circuits Syst. II* **53**(10), 1133–1137 (Oct 2006)
13. C. Vogel, D. Draxelmayr, F. Kuttner, Compensation of timing mismatches in time-interleaved analog-to-digital converters through transfer characteristics tuning, in *Proceedings of 2004 Midwest Symposium on Circuits and Systems*, vol. 1 (July 2004), pp. I-341–344
14. Z. Liu, M. Furuta, S. Kawahito, Simultaneous compensation of RC mismatch and clock skew in time-interleaved S/H circuits. in *IEICE Trans. Electron.* **E89-C**(6), 710–716 (June 2006)
15. J.J.F. Rijns, H. Wallinga, Spectral analysis of double-sampling switched-capacitor filters. in *IEEE T. Circuits Syst.* **38**(11), 1269–1279 (Nov 1991)
16. N. Kurosawa et al., Explicit analysis of channel mismatch effects in time-interleaved ADC systems. in *IEEE T. Circuits Syst. I* **48**(3), 261–271 (March 2001)
17. L. Xiaopeng, A.R. Bugeja, M. Ismail, A fast and accurate calibration method for high-speed high-resolution pipeline ADC, in *IEEE International Symposium on Circuits and Systems (ISCAS'02)*, vol. 2 (May 2002), pp. 26–29
18. F. Daihong, K.C. Dyer, S.H. Lewis, P.J. Hurst, A digital background calibration technique for time-interleaved analog-to-digital converters. in *IEEE J. Solid-St. Circ.* **33**(12), 1904–1911 (Dec 1998)
19. K.C. Dyer, F. Daihong, S.H. Lewis, P.J. Hurst, An analog background calibration technique for time-interleaved analog-to-digital converters. in *IEEE J. Solid-St. Circ.* **33**(12), 1912–1919 (Dec 1998)
20. Y.C. Jenq, Direct digital synthesizer with jittered clock. in *IEEE Trans. Instrum. Meas.* **46**(3), 653–655 (June 1997)
21. C.-C. Hsu, F.-C. Huang, C.-Y. Shih, C.-C. Huang, Y.-H. Lin, C.-C. Lee, B. Razavi, An 11b 800MS/s time-interleaved ADC with digital background calibration, in *ISSCC Digest of Technical Papers* (Feb 2007), pp. 464–615
22. S. Huang, B.C. Levy, Adaptive blind calibration of timing offset and gain mismatch for two-channel time-interleaved ADCs. in *IEEE Trans. Circuits Syst. I* **53**, 1278–1288 (June 2006)
23. J.E. Freund, *Mathematical Statistics* (Prentice Hall, Englewood Cliffs, NJ, 1992)

Chapter 5
Design of a 1.2 V, 10-bit, 60–360 MHz Time-Interleaved Pipelined ADC

5.1 Introduction

ADCs with resolution of eight to 12 bits and sampling rates of several hundreds of mega-samples-per-second (MS/s) found increasing importance in various types of wide-band applications, like in wireless transceivers, medical imaging, video signal processing, data acquisition and instrumentation. Very often the limitations in achieving such speed ranges of the ADC's specifications are one of the whole system bottlenecks. The possibility of having reconfigurable speed/power options in a single ADC design is also important for various applications that require varying speed, like in multi-standard transceivers or in various video with different resolutions and refresh rate. This characteristic permits to obtain a good compromise in terms of speed-per-power over various speed requirements [1].

This chapter will present the implementation details of a 1.2 V, 10-bit, 60–360 MS/s time-interleaved reset-opamp pipelined ADC implemented in 0.18 μm CMOS technology with Metal-Insulator-Metal (MiM) capacitor option, and for the mid-supply floating switches, the threshold voltages are $V_{thn}/V_{thp} = 0.63$ V/–0.65 V with the consideration of body effects, thus the sum of threshold voltages are greater than the supply voltage so that floating switches cannot be utilized. The key features employed in the ADC include:

1. Six time-interleaved channels applying a *front-end resistive-demultiplexing technique* implemented in a low-voltage environment.
2. All the low-voltage circuits techniques presented in Chapter 3 to achieve high-speed SC pipelined ADC implementation.
3. Several speed options, namely 60, 120, 180, 240 and 360 MS/s with corresponding scaled power consumption obtained through the selection of different number of channels that can be either 1, 2, 3, 4 or 6, respectively.
4. Low-voltage current-mode sub-ADCs which allow current-mirror sharing between the low-voltage comparators thus implying a reduction in static power consumption.

S.-W. Sin et al., *Generalized Low-Voltage Circuit Techniques for Very High-Speed Time-Interleaved Analog-to-Digital Converters*, Analog Circuits and Signal Processing, DOI 10.1007/978-90-481-9710-1_5, © Springer Science+Business Media B.V. 2010

5. A programmable timing-skew insensitive clock generator that alleviates the timing-mismatches among different channels and at the same time provides programmable multi-phase clocks for different speed options. Digital calibration usually reveals itself as a much more complex and costly solution to calibrate the dynamic timing-mismatches, especially in the 0.18 μm CMOS technology node.

All the techniques mentioned above are implemented without using any on-chip high-voltage to alleviate floating switch problems. On the other hand, they also exhibit a good potential to be applied in more advanced technology nodes (obtained through continuing deep-submicron CMOS down-scaling), as it is here clearly demonstrated, in particular by the very small headroom between the threshold and supply voltages.

5.2 The Overall ADC Architecture

Figure 5.1 shows the functional block diagram of the proposed ADC architecture that is composed by six time-interleaved channels. Each channel comprises a 60 MS/s pipelined ADC, as shown in Fig. 5.2, with a front-end S/H and eight 1.5 b MDAC stages, as well as a final 2 b flash ADC, like in typical 1.5 b/stage architectures. The structure includes built-in Digital Error Correction (DEC) logic in each channel producing a 10 b digital code that can be eventually combined with the output digital multiplexer. To implement input demultiplexing in a low-voltage environment, a

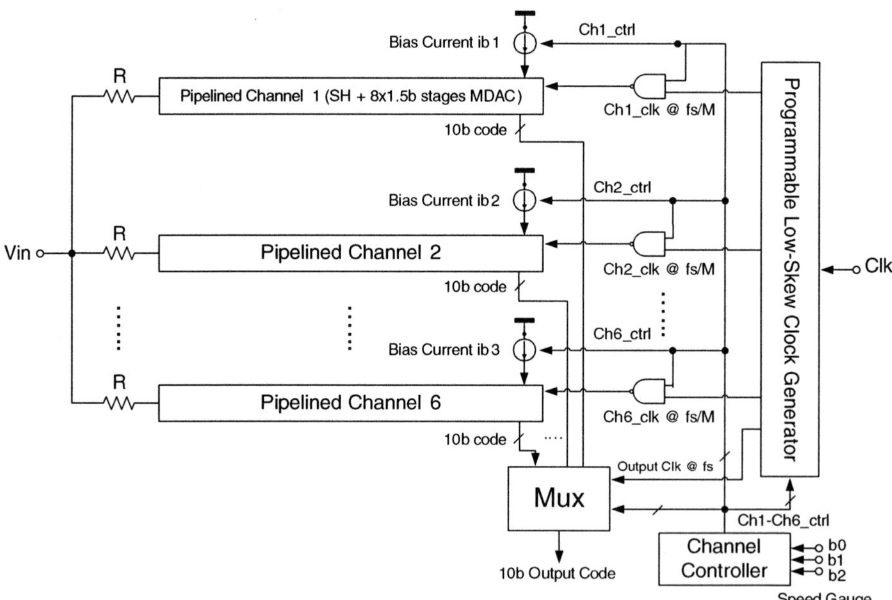

Fig. 5.1 Proposed six channels time-interleaved ADC architecture with speed-scalable options

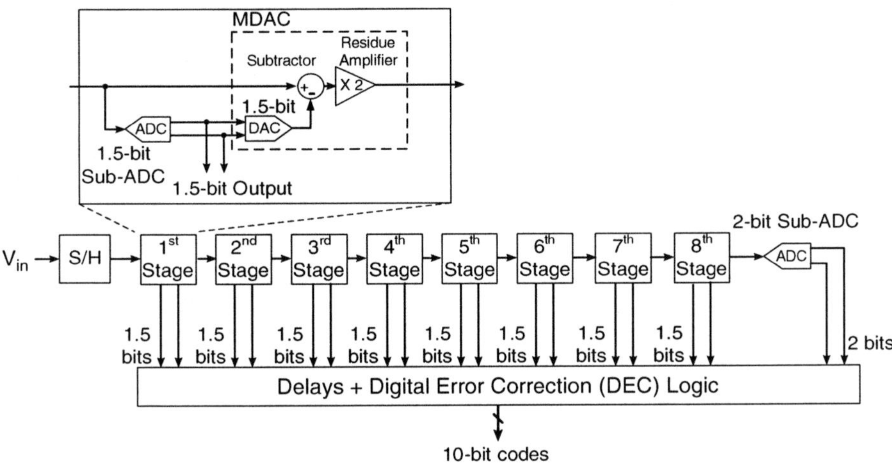

Fig. 5.2 The pipelined ADC architecture in one channel

resistor R is used (before each channel) to interface with the input signal. This, associated with the crossed-coupled S/H (described in next section) will allow demultiplexing the input signal into each channel without signal feedthrough problems.

The power/speed scalable option is implemented by selectively powering-down the unused channels (by switching-off the biasing currents in the opamps) since in a time-interleaved ADC the overall conversion speed is proportional to the number of channels being used. As depicted in Fig. 5.1, a channel controller is used to activate/ deactivate the bias currents of the opamps in various channels, and NAND gates are also utilized before the time-interleaved clocks arrive to the gates of the analog switches in each individual channel to avoid switching operation during power-down. The programmable clock generator (controlled by the channel controller) automatically provides the corresponding clock phases for different speed options, and the 10 b codes from various channels are combined in the final Multiplexer (Mux) to form a single set of 10 b output codes. The method described achieves very high-speed of operation with power/speed scaling options due to the high-speed nature of time-interleaved ADCs.

5.3 Prototype Circuit-Level Design

5.3.1 *Resistively Demultiplexed Front-End Sample-and-Hold*

Time-interleaved techniques are difficult to be applied in low-voltage environments due to the inherent problems of floating switches [2–4]. Here, a resistive demultiplexing technique is proposed which, if combined with a crossed-coupled reset-opamp S/H as presented in Chapter 3, can provide input demultiplexing without

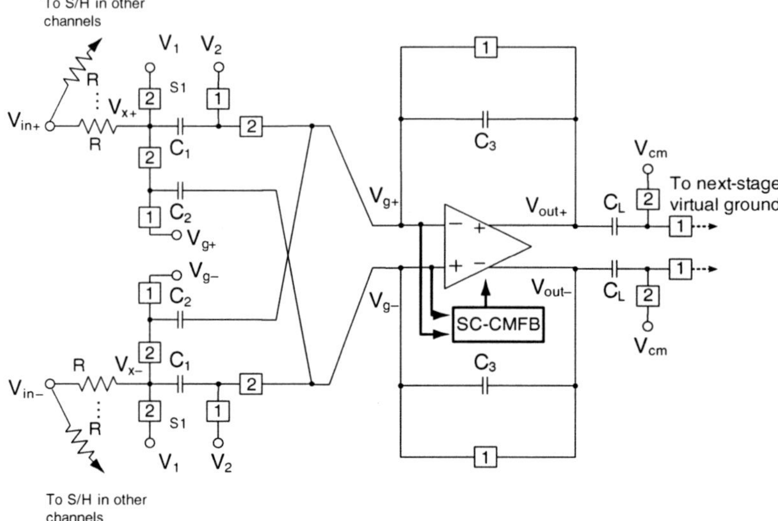

Fig. 5.3 Crossed-coupled S/H with resistive front-end demultiplexing

direct signal-feedthrough problems, as illustrated in Fig. 5.3. Signal-feedthrough is
generated (although attenuated) which can be cancelled by the cross-connected
capacitor C_2. As a result the input signal is decoupled between different channels
thus allowing the demultiplexing operation.

The output common-mode voltage of the opamp in Fig. 5.3 should be set at mid-
supply in amplification phase to maximize the output swing, and the VCLS
techniques in Section 3.4 is utilized here. $C_1 = C_2 = C_3$ to implement a gain of 1,
and V_1, V_4 are fixed potentials used to obtain the required level-shifting function.
For this S/H circuit the following relationship holds:

$$
V_{out,CM}[\phi 2] = \frac{C_1}{C_3} V_{in,CM} + \left(1 + \frac{C_1 + C_2}{C_3}\right) V_{G,CM}
$$
$$
- \frac{1}{C_3}[C_1 V_2 + (C_1 + C_2)V_1]
$$

(5.1)

which shows large flexibility for adjusting the value of the level shift. For $C_1 =
C_2 = C_3$, $V_{in,CM} = 0.6$ V and $V_{G,CM} = 0.75\, V_{DD} = 0.9$ V for suitable biasing point of
the NMOS differential pair of the opamp, the bias voltages can be simply chosen as
$V_1 = V_2 = 0.9$ V such that $V_{out,CM}[\phi 2] = 0.6$ V.

5.3.2 *1.5 b/Stage Multiplying-Digital-to-Analog-Converter (MDAC)*

Figure 5.4 shows the circuit diagram of the 1.5 b MDAC stage used in the pipelined
ADC. The MDAC is gain-and-offset compensated using the techniques in Chapter 3.

Only the five stages of the front-end MDACs have utilized the auxiliary amplifier to compensate the gain error. All the main opamps use the current-mirror topologies as shown in Fig. 3.12, while the auxiliary opamps utilize the differential-difference opamps as shown in Fig. 3.15, with the device size of the first stage main and auxiliary opamp presented in Table 5.1. The sampling capacitors and the opamps are scaled down along the pipelined stages to save power.

The implementation of the reference injection circuit is also shown in Fig. 5.4. Unlike the traditional flip-over MDAC architecture [5] that can achieve the most excellent feedback factor of 0.5, in low-voltage designs this architecture cannot be utilized due to the floating switches problems. Extra capacitors C_{ref} are used in Fig. 5.4 to inject a reference-dependent charges into the virtual ground of the opamp to achieve the reference subtraction, and the output voltage of the MDAC can be expressed as (considered only the differential mode):

$$V_{o1} - V_{o2} = \frac{C_{s1}}{C_{f1}} V_{in} - m V_{DD} \frac{C_{ref}}{C_{f1}} + \text{Gain error} \tag{5.2}$$

where m would be equal to either 1, 0, or -1 depending on the sub-ADC decision (corresponding to code 10, 01 and 00 in Fig. 5.4 respectively). The amount of reference voltage injected is defined as a ratio of supply voltage, and $C_{ref} : C_{f1} : C_{s1} = 1 : 2 : 4$ is used to implement the desired reference voltage injection and the 2x MDAC gain, with the feedback factor reduced to 0.28 even without counting the parasitic input capacitances of the opamps.

Notice that two separated reference capacitors are used to handle the different common-mode injection requirements through the reference capacitors depending on the sub-ADC decision. When m is equal to 1 (or -1), the left side of the capacitor C_{ref} is connected to V_{DD} (or GND) while in the opposite differential path it is connected to GND (or V_{DD}). This is equivalent to connect a mid-supply common-mode voltage during the code decision $m = 1$ or -1, but for the case of $m = 0$, no reference charges are needed to be injected and thus the mid-supply common-mode voltage cannot be produced due to the lack of floating switches. In order to ensure equal-common-mode voltage injection among different codes, different C_{ref} is utilized for different code decision in sub-ADC. It can be deduced the following CM relationship (assuming that both the main and auxiliary opamps

Table 5.1 Device size for main and auxiliary opamp in first MDAC

Transistor	Size (mm)	Transistor	Size (mm)
M0	12/1 × 64	M0A, M0B	12/1 × 64
M1A, M1B	6/0.25 × 32	M1A–M1D	6/0.25 × 32
M2A, M2B	3.5/0.25 × 84	M2A, M2B	3.5/0.25 × 74
M3A, M3C	2.5/0.35 × 4	M3A, M3C	2.5/0.35 × 4
M3B, M3D	2.5/0.35 × 36	M3B, M3D	2.5/0.35 × 16
M4A, M4B	17.5/0.18 × 90	M4A, M4B	17.5/0.18 × 40
M5A, M5B	16/0.18 × 90	M5A, M5B	16/0.18 × 40

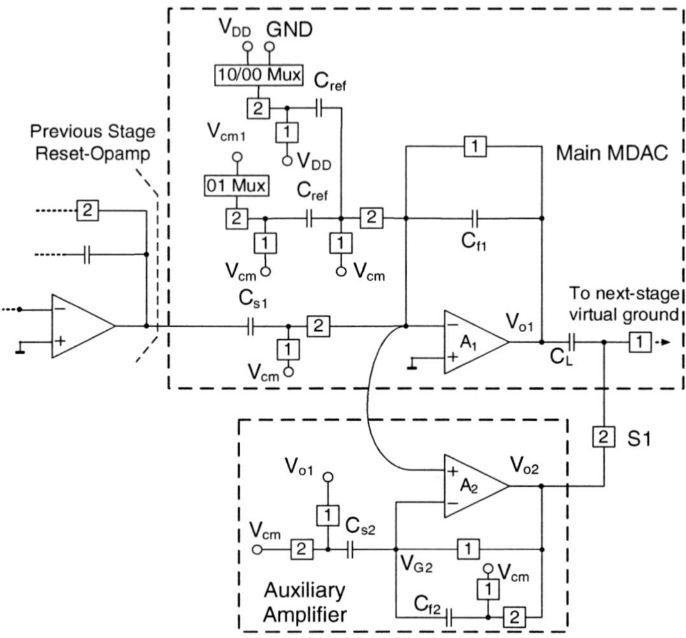

Fig. 5.4 A reset-opamp MDAC with gain-and-offset compensation

in all stages have the same $V_{G,CM} = V_{CM} = 0.9$ V and considering the previous stage reset to $V_{G,CM}$ in phase 2):

$$V_{o1,CM}[\phi 2] = \frac{C_{s1}}{C_{f1}} V_{in,CM} + \frac{C_{ref}}{C_{f1}} \left(\frac{1}{2} V_{DD} \right) + \left(1 - \frac{C_{s1}}{C_{f1}} \right) V_{CM} \text{ for } m = 1 \text{ or} - 1 \quad (5.3)$$

$$V_{o1,CM}[\phi 2] = \frac{C_{s1}}{C_{f1}} V_{in,CM} - \frac{C_{ref}}{C_{f1}} V_{CM1} + \left(1 + \frac{C_{ref} - C_{s1}}{C_{f1}} \right) V_{CM} \text{ for } m = 0 \quad (5.4)$$

By choosing $V_{CM} = 0.9$ V, $V_{CM1} = 0.3$ V, $V_{in,CM} = 0.6$ V then both (5.3) and (5.4) evaluate to 0.6 V. Actually to achieve the require level shifting, only two extra common-mode voltage ($V_{CM} = 0.9$ V, $V_{CM1} = 0.3$ V) is required. Since the common-mode voltage doesn't require being accurate, these CM voltages can be generated by simple resistive ladder with bypassing capacitors.

5.3.3 Current-Mode Sub-ADC Design

The 1.5 b sub-ADC contains two comparators while the final 2 b sub-ADC contains three. Conventional voltage-mode differential pair architectures cannot be used in

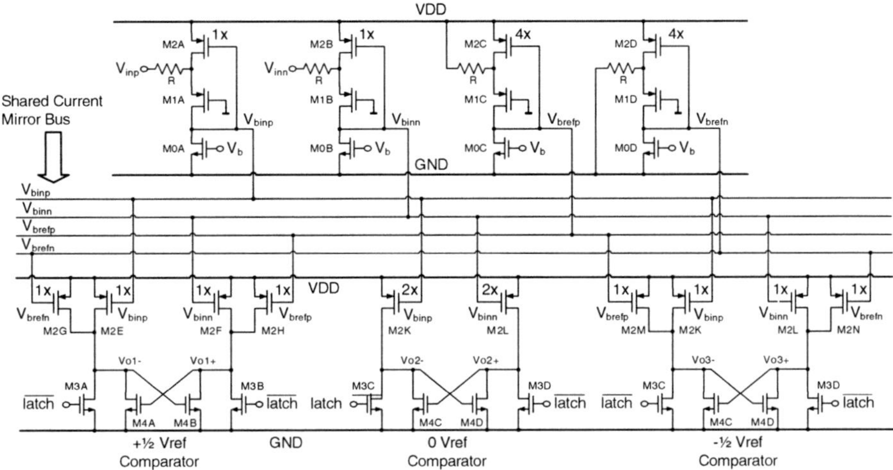

Fig. 5.5 The low-voltage current-mode 2 b sub-ADC architecture

low-voltage designs due to large voltage swing and input common-mode range. Also, switched-capacitor comparators cannot be utilized due to the floating switches. A current-mode comparator [6] will be adopted here based on its low-voltage operation capability and adjustable threshold. It will also exhibit very low kick-back noise from the latch since it is shielded in two steps by the current-mirror and the input resistors. However, due to the operation in current-mode the comparators will draw static power in the current mirrors. To reduce the static power a low-voltage current-mode sub-ADC architecture is proposed, as shown in Fig. 5.5 in its 2 b version, to share the static current mirror in the sub-ADC. The current mirror voltages V_{binp}, V_{binn} are generated from the input signal V_{inp} and V_{inn}, and on the other hand V_{brefp}, V_{brefn} are generated from the supply rails to produce reference voltages for the comparators. These mirror nodes can be shared among three comparators (Fig. 5.5) since two of three comparators have identical thresholds with different polarities ($+/-0.5$ V_{ref} comparator thresholds as required in 2 b stages), while the zero-reference comparator does not require reference voltages. The zero-crossing points of the $+/-0.5$ V_{ref} comparators are defined as follows:

$$V_{inp} - V_{inn} = \pm\tfrac{1}{4}V_{DD} = \pm\tfrac{1}{2}V_{ref} \tag{5.5}$$

5.3.4 Programmable Timing-Skew Insensitive Clock Generator

Besides offset- and gain-mismatches, sampling-time mismatches in various time-interleaved channels also create modulated sidebands which will degrade the performance of the ADC [7]. A low-skew clock generator (shown in Fig. 5.6) is

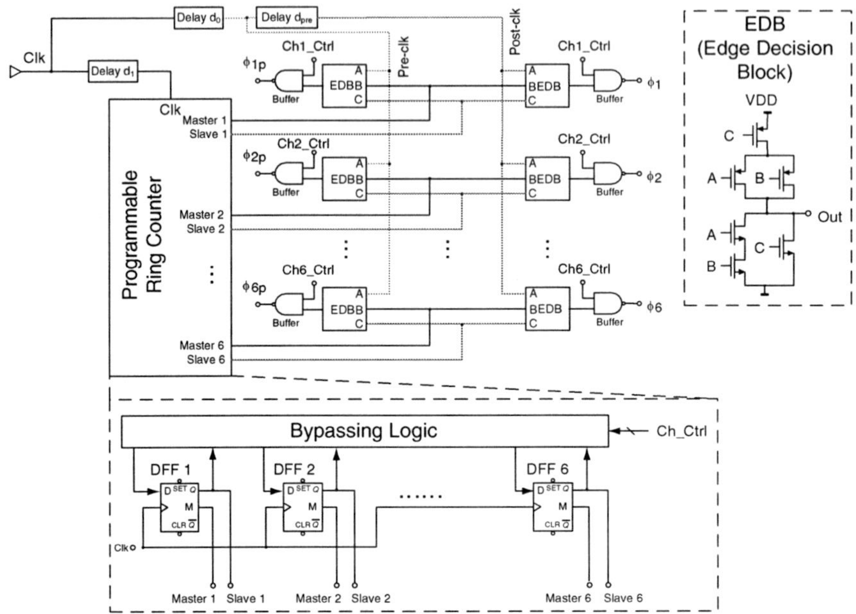

Fig. 5.6 Programmable low-skew clock generator

used here to reduce the sensitivity due to the timing-skew effect, which relies on assigning the common master Pre-clk signal for the decision of the sampling instant. This is a modified version from [8] that provides correct clock phases for speed scaling options in which some of the D-Flip-Flops in the ring-counter are bypassed depending on the channel controller output (Fig. 5.6).

5.3.5 Noise Analysis

Circuit noise is one of the main dominant limiting factors to the achievable dynamic range of the ADC, especially in low-voltage circuit which has intrinsically reduced signal swing and thus the signal power, as well as the added extra circuitries (and extra noise) in various low-voltage circuit solutions, e.g. reset-opamp technique and auxiliary gain compensation. Circuit noises are originated from random fluctuations due to on-resistance of the switches as well as the transistors in the opamps, and flicker noise is negligible compared with thermal noise contribution in broadband sampled-data systems [9].

In SC circuit when the input signal is passively sampled into the capacitor, the white noise spectrum generated by the channel on-resistance of the switches will be filtered by the first order RC low-pass transfer function. Since the same resistance R is used in generation of thermal noise as well as the bandwidth, the total sampled noise power in the capacitor (which is typical in sampling phase) is governed by the well-know kT/C noise. However during the amplification phase the switches are used to discharge the sampling capacitor into the virtual ground of the opamps and in this case the white noise of the switch's resistance is shaped by the opamp's closed-loop bandwidth which is predominant one. In this case, the resulted output noise of the SC stage will also depends on the channel resistance, which involves more complex manipulations.

As an example, consider the crossed-coupled S/H circuit as shown in Fig. 5.3. The thermal noise contributions from both the switches and opamp in the sampling and reset phase can be represented in Fig. 5.7. As shown in Appendix C, it can be shown that the output noise of the S/H can be evaluated as

$$\overline{v^2_{n,S/H}} = 2\left\{ 2kT/C + \omega_{GBW\_\phi1}\left[kTR_{s2,A} + \frac{1}{2}S_{opamp}(0)\right] \right.$$

$$\left. + \omega_{GBW\_\phi2}\left[\frac{1}{4\beta}S_{opamp}(0) + kT\beta(R_{s1,B} + R_{s2,B})\right] \right\} \qquad (5.6)$$

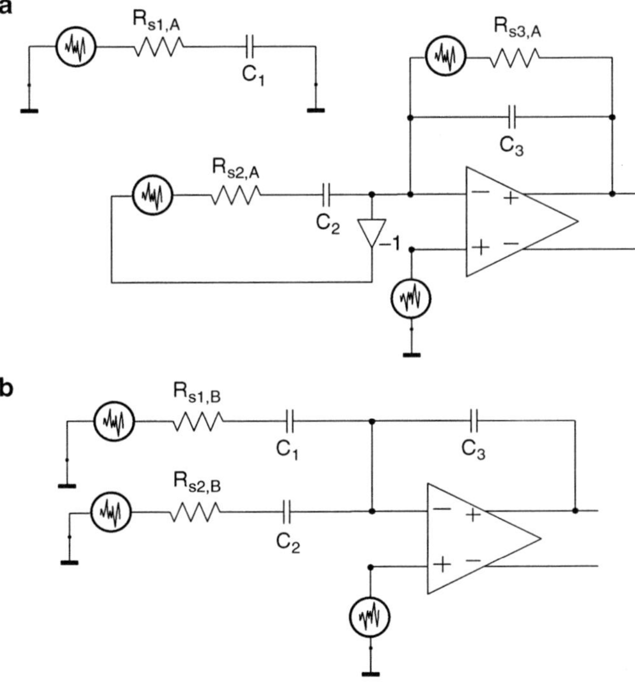

Fig. 5.7 Noise in the crossed-coupled S/H for (a) sampling phase and (b) amplification phase

where $\omega_{GBW\_\phi m}$ is the gain-bandwidth of the opamp in phase m, $S_{opamp}(0)$ is the input referred noise Power Spectral Density (PSD) at low frequency, β is the feedback factor and all resistances are represented in Fig. 5.7. Similarly for the MDAC equivalent circuit in Fig. 5.8, the noise can be evaluated as

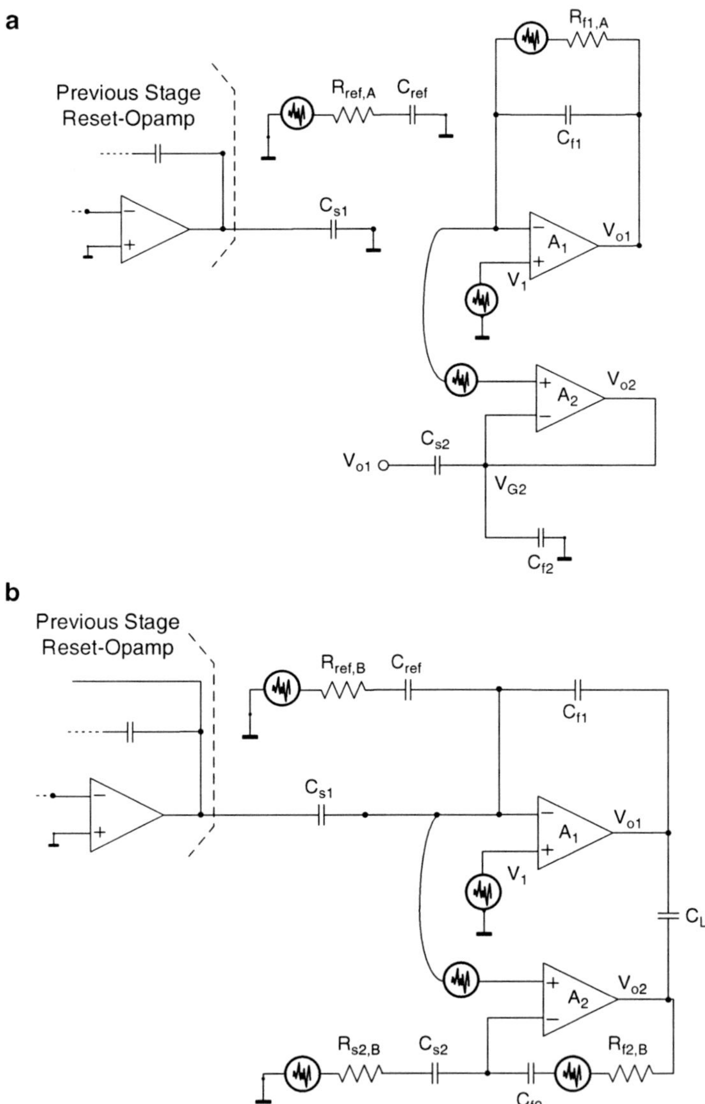

Fig. 5.8 Noise in the MDAC for (a) sampling phase and (b) amplification phase

$$\overline{v_{n,MDAC}^2[\phi 2]} = 2kT/C_{ref}\left(\frac{C_{ref}}{C_{f1}}\right)^2 + 2kT/C_{f1}$$

$$+ \frac{S_{main}(0)}{2}\left\{\frac{\left[\omega_{GBW\_main\_\phi 1}\left(\frac{C_{s2}}{C_{f2}}\right) + \sqrt{\omega_{GBW\_main\_\phi 1}\omega_{GBW\_aux\_\phi 1}}\right]^2}{\omega_{GBW\_main\_\phi 1} + \omega_{GBW\_aux\_\phi 1}}\right.$$

$$\left. + \frac{\omega_{GBW\_main\_\phi 2}^2/\beta}{\omega_{GBW\_main\_\phi 2} + \omega_{GBW\_aux\_\phi 2}}\right\}$$

$$+ \frac{1}{2}S_{aux}(0)\left\{\omega_{GBW\_aux\_\phi 1}\left[1 + \frac{C_{s2}}{C_{f2}}\right]^2 + \frac{1}{\beta}\omega_{GBW\_aux\_\phi 2}\right\}$$

$$+ \frac{1}{2}S_{pre}(0)\left(\frac{C_{s1}}{C_{f1}}\right)^2\beta\omega_{GBW\_main\_\phi 2}$$

$$+ 2kT\beta\left\{R_{ref,B}\left(\frac{C_{ref}}{C_{f1}}\right)^2\omega_{GBW\_main\_\phi 2}\right.$$

$$\left. + \left[R_{s2,B}\left(\frac{C_{s2}}{C_{f2}}\right)^2 + R_{f2,B}\right]\omega_{GBW\_aux\_\phi 2}\right\} \tag{5.7}$$

where $\omega_{GBW\_main\_\phi m}$ and $\omega_{GBW\_aux\_\phi m}$ is the gain-bandwidth of the main and auxiliary opamp in phase m, $S_{opamp}(0)$ and $S_{aux}(0)$ is the input referred noise PSD of both opamp at low frequency, respectively, and $S_{pre}(0)$ is the input referred noise PSD of previous stage's opamp at low frequency.

According to on-resistance value and the input referred PSDs of the opamp as obtained by transistor simulations, the input referred noise voltage of the pipelined ADC can be calculated as 484 μV_{rms}, compared with the noise simulated using Transient Noise analysis in Cadence which is 522 μV_{rms}. Together with the quantization noise this will limit the SNR of the whole ADC to 56.6 dB.

5.3.6 *Channel Mismatch Analysis*

Since the ADC is time-interleaved in six channels, channel mismatches should be characterized. Table 5.2 shows the contribution of the various mismatch parameters and the resulting SNDR from the formula derived in Chapter 4, including noise budget as presented before. It can be seen that the performance of the ADC is limited by noise.

Table 5.2 Contribution of various mismatches and noise error sources

	Offset (V)	Gain	Timing (s)	Bandwidth	Noise	Total
Sigma	5.00E-04	8.00E-04	2.00E-12	2.00E-03		
SNDR limit	59.4	62.7	66.8	63.3	56.6	53.4

5.4 Layout Considerations

This high-speed ADC is laid out in a 0.18 μm single-poly, six-metal CMOS process with MiM capacitor options. Considering the highest clock frequency inside the chip is running at 360 MHz and various signals and clocks are required to distribute to various time-interleaved channels, special attention was paid to the design of the layout, including device and path matching, the parasitic RC-delay in the metal routings, shielding of sensitive signal (such as virtual grounds), isolation between analog, digital as well as IO data buffer domains, and power-ground distribution network.

Figure 5.9 shows the floor plan and the die microphotograph of the prototype ADC. The total active area is 13.2 mm^2 (2.2 mm^2 per channel), which is dominated mainly by the large sizes of capacitors as well as the transistors in the opamps due to the reduced headroom for thermal noise and transistors' overdrive voltage. The six channels are laid out horizontally, and the channels are arranged such that, e.g. channel 2 and 5 share the same clock buses and thus are put adjacent to each other, since they share the same set of interleaved clocks (i.e. phase 1 in channel 2 is identical to phase 2 in channel 5).

By this specific layout placement of channels that sharing the same clock buses, the opamps between channels are placed in closer proximity such that their matching is better; However for the channels that are located far away (like channel 1 and 6), the matching of transistor between them is quite poor. This would induce large mismatch in the channel gain among them, especially for the front-end S/H. Such gain mismatch will be calibrated by digital post-processing as demonstrated in Chapter 1.

Special attention was also devoted to the supply bus routing due to the high-speed nature of the ADC chip. To minimize the supply impedance, power buses are routed with large sheets of parallel metal layers, with distributed PMOS decoupling capacitors [10–13] throughout the empty spaces of the chip, and multiples bondwires connected to the package to reduce the equivalent supply inductance which is one of the dominant causes of supply and substrate noise coupling [10, 14–19]. Furthermore, since inside the MDAC the supply voltages have been used to generate reference voltages, reference power buses are routed apart from the main power bus to avoid the resistive voltage-drops developed by the large static opamp currents, which cause code-dependent errors and lead to the degradation of ADC's linearity.

The analog input signal enters from the left side of the chip and then is routed to the interleaved S/H in each channel. This routing is carefully matched to minimize the bandwidth mismatch effect due to the additional parasitic RC routing delay. The clock generator and the digital circuits are located at the right side of the chip which is far away from the most sensitive front-end S/H and MDACs in the left side, to avoid noise

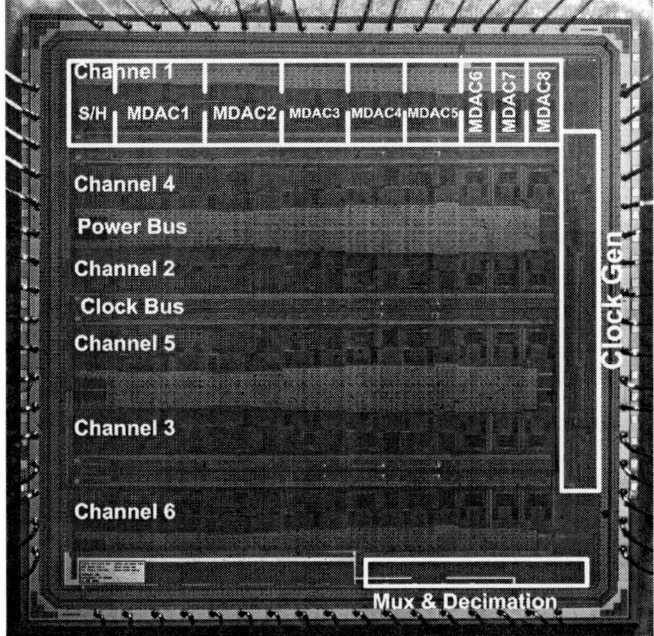

Fig. 5.9 Die microphotograph

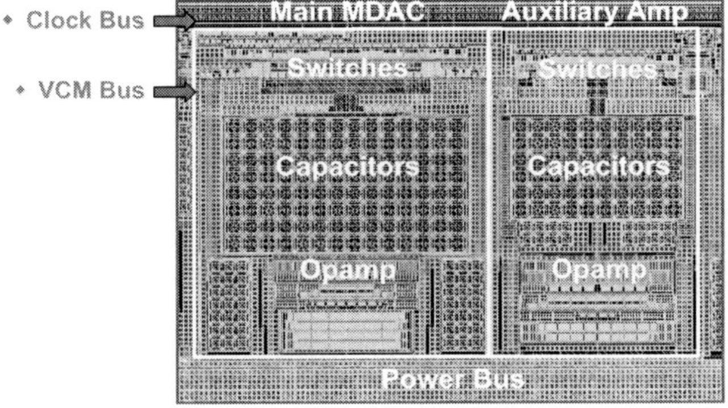

Fig. 5.10 The layout of the first stage MDAC

coupling from noisy digital substrate. The clock buses are also carefully matched to evenly distribute the time-interleaved clocks to the individual channel. The 10 b codes from each channel are then routed to the output digital multiplexer and decimator and then eventually output from the IO buffers from the bottom side of the chip.

Figure 5.10 shows the layout of the first stage MDAC, which is composed of the main MDAC plus the auxiliary amplifier for gain compensation. The opamps used in both amplifiers are laid out in symmetrical ways, and the capacitors are placed in common-centroid manner and dummies exteriors to reduce the gradient- and periphery-induced capacitor mismatches [20–26]. Sensitive signal such as virtual grounds are shielded using analog ground to avoid noise coupling which is crucial especially in high-speed mixed-signal circuit designs.

5.5 Simulation Results

5.5.1 Opamp Simulations

The speed of the opamps used in the pipelined ADC are crucial to the performance of the whole pipelined ADC linearity. Since the opamps are used in a negative-feedback configuration, stability must be ensured while maintain sufficient gain-bandwidth product to achieve the required settling accuracy. It is difficult to be achieved especially in reset-opamp circuits, since they are enforced to reset in resetting phase (feedback factor $\beta = 1$) for its basic operation, while in the amplification phase the feedback factor drops dramatically to $\beta = 0.23$ (including post-layout parasitics), as shown in Fig. 5.4. Such difference in feedback factor in both phases creates challenges in optimizing the Unity-Gain Frequency (UGF) and the phase margin of the opamp, since the non-dominant poles must be located at frequencies high enough so that the opamps are still stable in resetting phase, while it is not desire to over-compensate the opamp during the amplification phase which would affect its settling time. For the most stringent requirement on the first stage MDAC, AC open-loop post-layout simulations show that the main opamp in this MDAC dissipates around 7.6 mW under 1.2 V supply voltage, with 44 dB of DC-gain, 732 MHz of UGF, 29° of phase margin in resetting phase (as shown in Fig. 5.11a with corner cases), and 30 dB of closed-loop DC-gain, 316 MHz of closed-loop bandwidth, 81° of phase margin simulated at the virtual grounds of the opamp in the amplification phase (Fig. 5.11b with corner cases). Notice that although the phase margin of the opamp reset phase is as low as 30°, the opamp is still stable throughout the corners, and the opamp can still settle due to the large bandwidth in the reset-phase. Figure 5.12 also demonstrated the corresponding post-layout loop-gain simulation results for first stage auxiliary amplifier, showing that it achieve a 39 dB of closed-loop DC-gain, 175 MHz of closed-loop bandwidth, 72° of phase margin and consuming 6.7 mW. According to the gain-compensation scheme in Chapter 3, the equivalent opamp gain is equal to 83 dB which is sufficient for the first stage MDAC.

5.5.2 Front-End S/H

The front-end S/H has utilized the VG-CMFB technique as presented Chapter 3 to stabilize the output common-mode level, and together with the VCLS technique the output common-mode can be set at different voltage at different phase. Figure 5.13

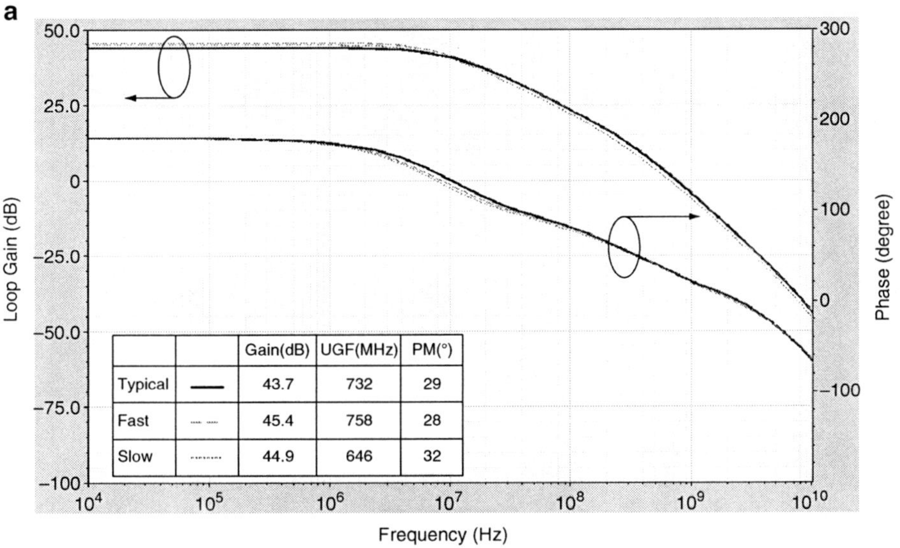

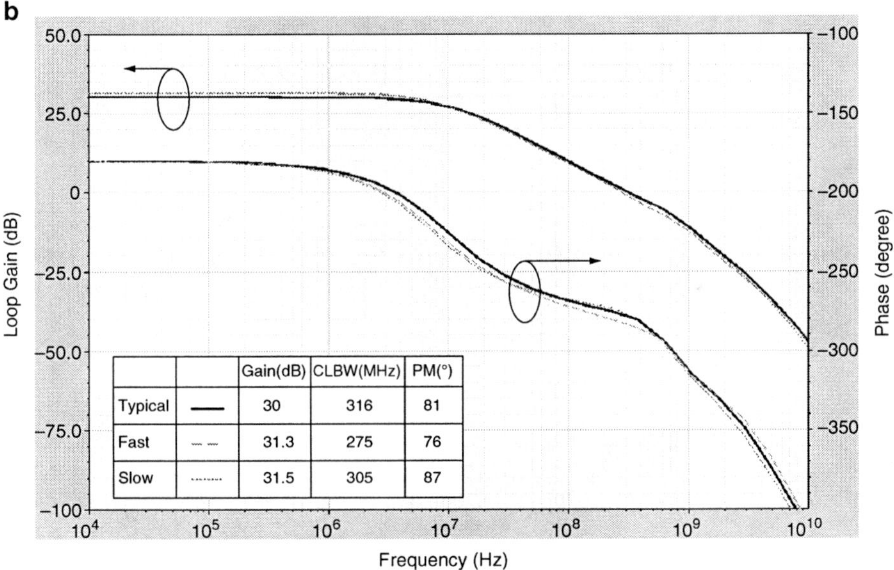

Fig. 5.11 Post-Layout loop-gain frequency response of the first MDAC's main opamp in (**a**) resetting phase and (**b**) amplification phase

shows the transient simulation waveforms, in which the single-ended output voltage and differential output are presented in Fig. 5.13a and b respectively. The S/H achieve a good linearity (<-65 dB Total Harmonic Distortion, THD), since no floating switches are presented and thus the S/H are inherently free of signal-dependent charge-injection and on-resistance from those switches.

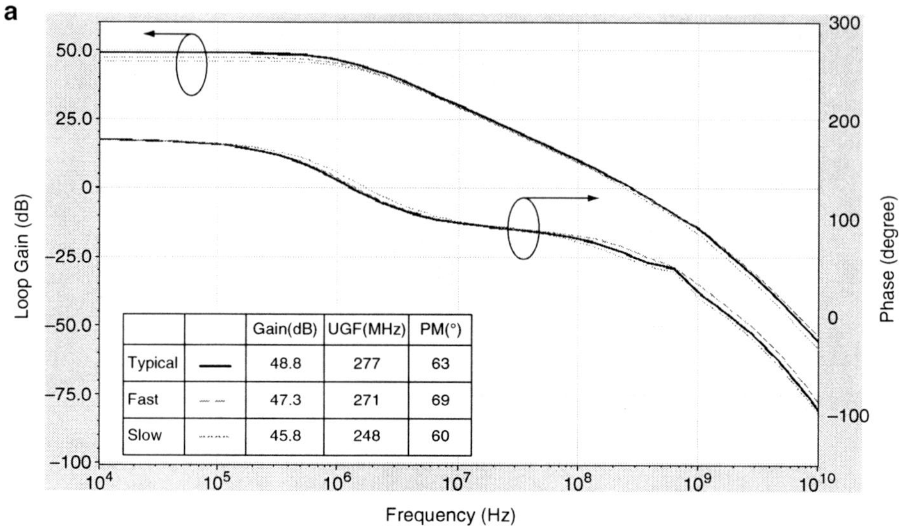

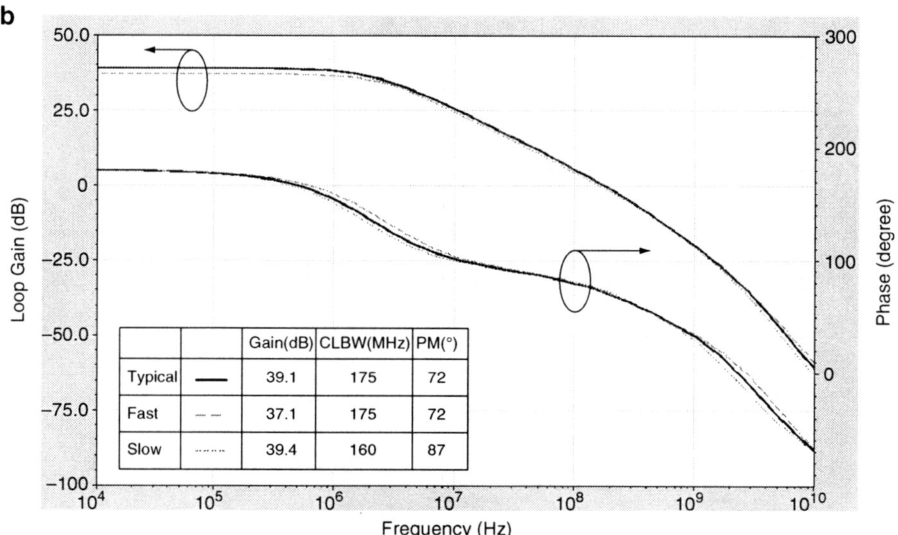

Fig. 5.12 Post-Layout loop-gain frequency response of the first MDAC's auxiliary opamp in (**a**) resetting phase and (**b**) amplification phase

5.5.3　Current-Mode Comparator

In the operation of 1.5 b/stage pipelined ADC, the offset voltage of the 1.5 b sub-ADCs are greatly relaxed to within 1/4 of full reference voltage only due to the use of digital error correction (DEC) techniques, while for the last 2 b sub-ADC, the offset voltage should be within 0.5 LSB.

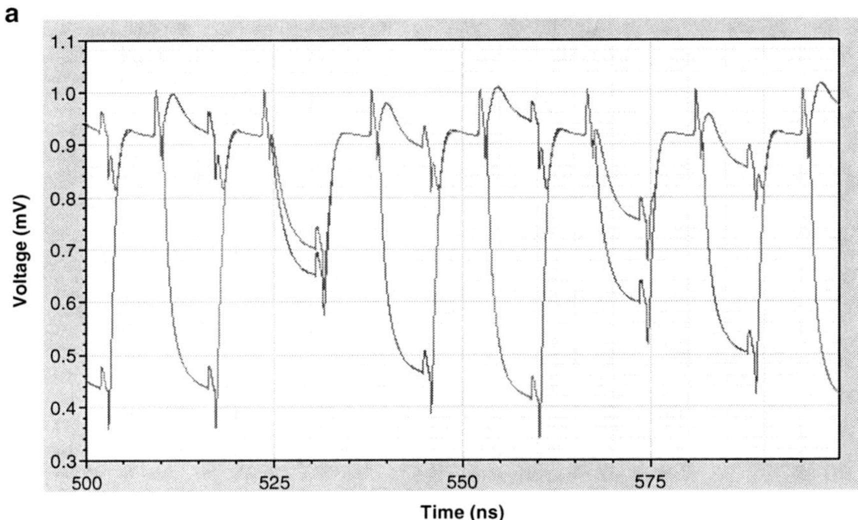

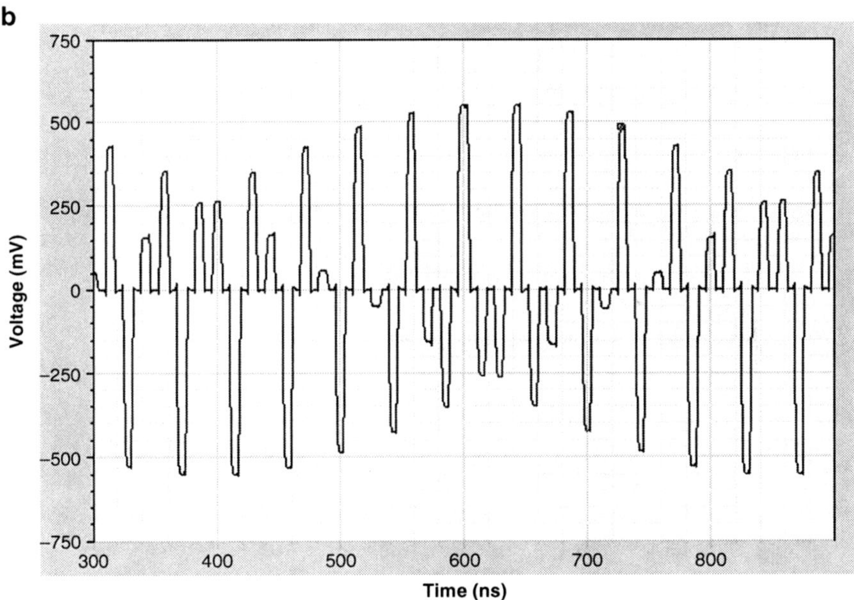

Fig. 5.13 Transient simulation of the S/H. (**a**) The single-ended output and (**b**) the differential-mode voltage

To ensure the offset voltage of all the current-mode sub-ADCs are within specification, post-layout Monte Carlo simulations are performed under various process corners, as shown in Fig. 5.14. The current consumption of the 1.5 b sub-ADCs are 300 μW each and the 2 b final flash consume totally 600 μW.

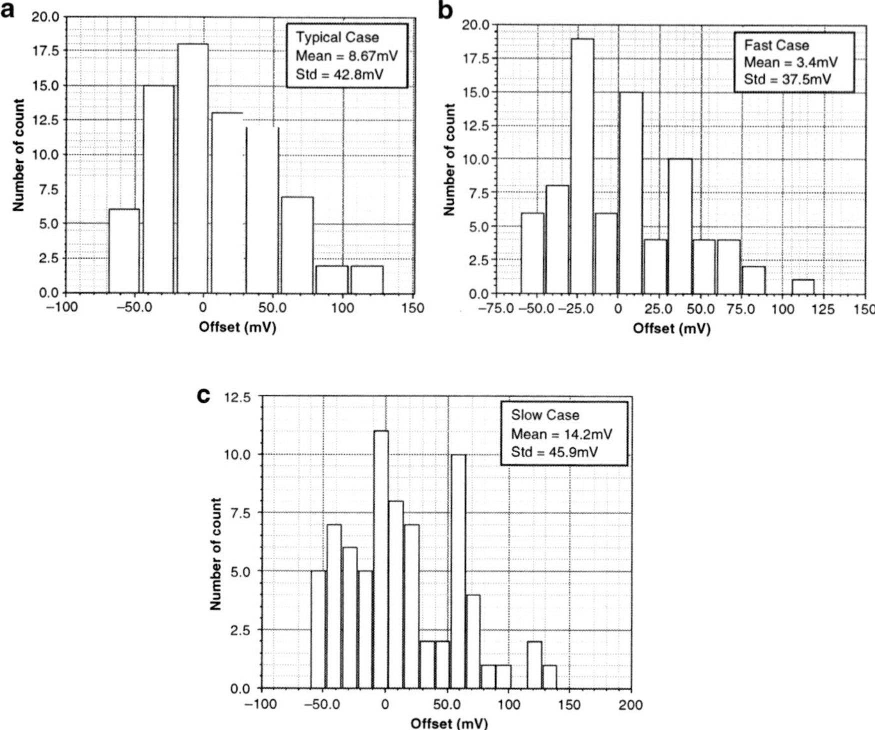

Fig. 5.14 Histogram of a 100-run Monte-Carlo simulation to mismatch-variation under (**a**) typical, (**b**) fast and (**c**) slow corner

5.5.4   The Overall ADC Simulations

The full chip was extracted and simulated with layout extraction and simulation package, e.g. spectre and spectreRF for simulations and Assura DRC, LVS and RCX for back-end verifications. Spectre transistor-level simulations are done with extracted layout parasitics including the bondwire inductance from all pins (e.g. supplies in analog and digital domain, clock and signal input, etc.). Channel mismatches (from MOSFETs, capacitors and resistors) are simulated through Monte-Carlo simulations. The SNDR includes the contributions of thermal and flicker noise which are simulated through the Transient Noise analysis from Cadence. The INL/DNL is too computational expensive to be simulated but will be measured in the real prototype as described in the test plan presented in the next section.

Figure 5.15 presents the time-domain reconstructed digital output codes of the 10 b ADC (1.2 Vpp sinusoidal input signal), with 1-ch operated @ 60 MS/s ($f_{in} = 24$ MHz) in Fig. 5.15a and 360 MS/s ($f_{in} = 37$ MHz) in Fig. 5.15b. Figure 5.16 shows the spectrum plot of one-case Monte-Carlo simulation results with $f_s = 360$ MS/s

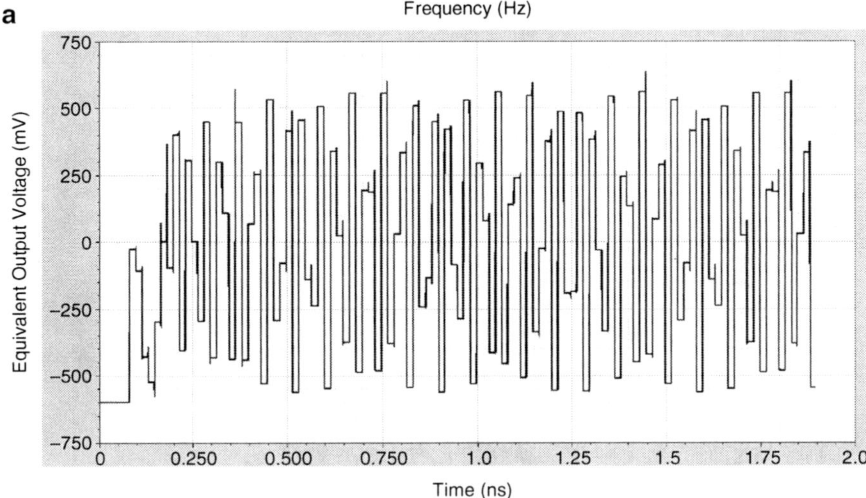

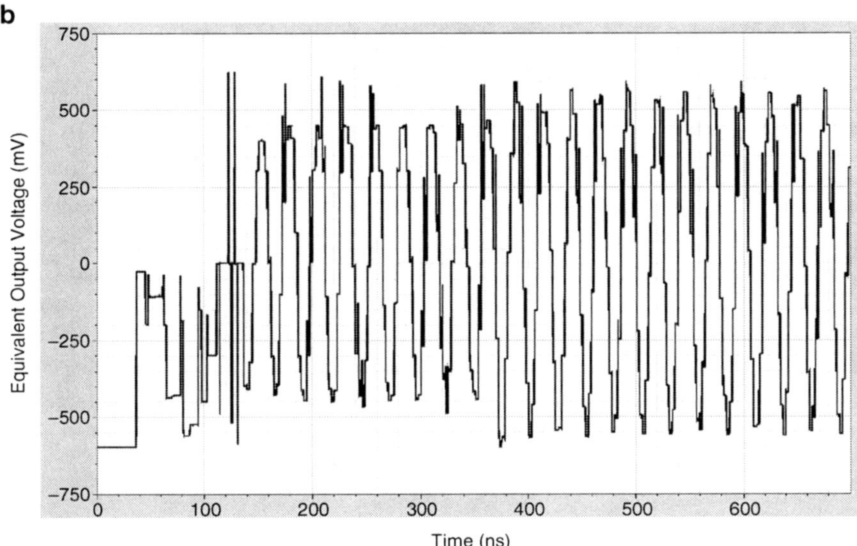

Fig. 5.15 Reconstructed ADC output code with (**a**) 1-ch 60 MS/s option and (**b**) 6-ch 360 MS/s option

at six channels and the spectrum clearly shows the offset-mismatch induced tone (represented as O) and gain-, timing- and bandwidth-mismatches induced tones (represented as G). The estimated offset in this case is $\sigma = 850\ \mu V$ according to the formula presented in Chapter 4, which is contributed mainly by the capacitor mismatch in the level-shifting circuits, since the input referred offset of the opamp in the front-end stage is as large as $\sigma > 3$ mV and be compensated by the offset cancellation scheme.

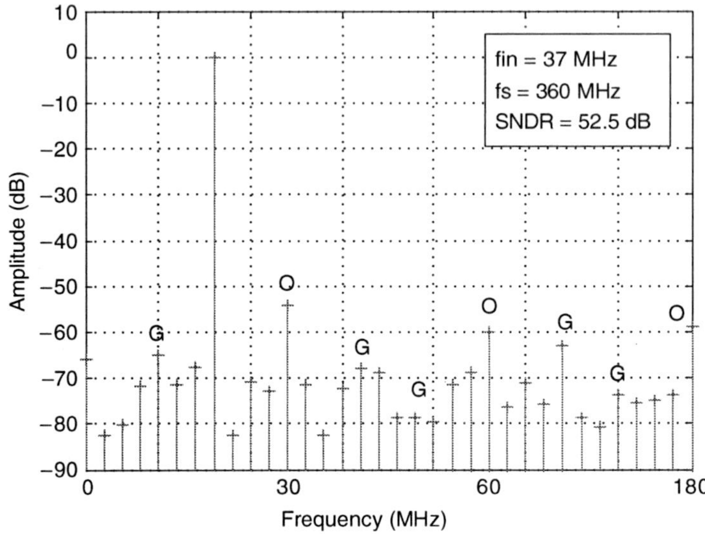

Fig. 5.16 Post-layout simulated output spectrum of the 6-ch TI-ADC

5.6 Summary

This chapter had presented the implementation details of a 1.2 V, 10 b, 60–360 MS/s reconfigurable time-interleaved pipelined ADC that employs all the proposed techniques presented in the book. The ADC is implemented in a 0.18 μm CMOS technology with MiM options ($V_{thn}/V_{thp} = 0.63$ V/−0.65 V for mid-supply floating switches). Practical prototype circuit implementation for each building block has been comprehensively investigated, with transistor-level as well as post-layout simulations. Simulation results show that the prototype achieves a satisfactory performance under process- and mismatch-variations.

References

1. I. Ahmed, D. Johns, A 50 MS/s (35 mW) to 1 kS/s (15 μW) power scalable 10b pipelined ADC with minimal bias current variation, in *Digest of Technical Papers of IEEE International Solid-State Circuits Conference (ISSCC)* (Feb 2005), pp. 280–281, 598
2. M. Keskin, U. Moon, G.C. Temes, A 1-V 10-MHz clock-rate 13-bit CMOS ΔΣ modulator using unity-gain-reset op amps. in *IEEE J. Solid St. Circ.* **37**(7), 817 (July 2002)
3. J. Li, G. Ahn, D. Chang, U. Moon, A 0.9-V 12-mW 5-MSPS algorithmic ADC with 77-dB SFDR. in *IEEE J. Solid St. Circ.* **40**(4), 960–969 (April 2005)
4. V.S.L. Cheung, H.C. Luong, W.H. Ki, A 1-V 10.7 MHz switched-opamp bandpass ΣΔ modulator using double-sampling finite-gain-compensation technique. in *IEEE J. Solid St. Circ.* **37**(10), 1215–1225 (Oct 2002)

5. S. Lewis, S. Fetterman, G. Gross, R. Ramachandran, T. Viswanathan, A 10-b 20-M samples/s analog-to-digital converter. in *IEEE J. Solid St. Circ.* **27**, 351–358 (March 1992)
6. S. Mortezapour, E.K.F. Lee, A 1-V, 8-bit successive approximation ADC in standard CMOS process. in *IEEE J. Solid St. Circ.* **35**(4), 642–646 (April 2000)
7. U. Seng-Pan, S.-W. Sin, R.P. Martins, Exact spectra analysis of sampled signal with jitter-induced nonuniformly holding effects. in *IEEE Trans. Instrum. Meas.* **53**(4), 1279–1288 (Aug 2004)
8. S.-W. Sin, U. Seng-Pan, R.P. Martins, A generalized timing-skew-free, multi-phase clock generation platform for parallel sampled-data systems. Proc ISCAS **1**, I-369–I-372 (May 2004)
9. C.A. Gobet, A. Knob, Noise analysis of Switched Capacitor networks. in *IEEE Trans. Circuit Syst.* **CAS-30**(1), 37–43 (Jan 1983)
10. AMS, Analog Group, *Crosstalk in Mixed-Signal Systems* (AMS, 1996)
11. D.K. Su, M.J. Loinaz, S. Masui, B.A. Wooley, Experimental results and modeling techniques for substrate noise in mixed-signal integrated circuits. in *IEEE J. Solid St. Circ.* **28**(4), 420–429 (April 1993)
12. L.K. Wang, H.H. Chen, On-chip decoupling capacitor design to reduce switching-noise-induced instability in CMOS/SOI VLSI, in *Proceedings of 1995 IEEE International SOI Conference* (Oct 1995), pp. 100–103
13. P. Larsson, Parasitic resistance in an MOS transistor used as on-chip decoupling capacitance. in *IEEE J. Solid St. Circ.* **32**(4), 574–576 (April 1997)
14. T.J. Schmerbeck, Noise coupling in mixed signal ASICs, Chapter 10, in *Low-Power HF Microelectronics: A Unified Approach*, ed. by G. Machado (IEE Press, New York, 1996)
15. T. Blalack, *Switching Noise in Mixed-Signal Integrated Circuits*, Ph.D. Dissertation, Stanford University, 1997
16. B.R. Stanisic, N.K. Verghese, R.A. Rutenbar, L.R. Carley, D.J. Allstot, Addressing substrate coupling in mixed-mode IC's: simulation and power distribution synthesis. in *IEEE J. Solid St. Circ.* **29**(3), 226–237 (March 1994)
17. X. Aragonès, A. Rubio, Experimental comparison of substrate noise coupling using different wafer types. in *IEEE J. Solid St. Circ.* **34**(10), 1405–1409 (Oct 1999)
18. Y. Zinzius, E. Lauwers, G. Gielen, W. Sansen, Evaluation of the substrate noise effect on analog circuits in mixed-signal designs, in *Proceedings of South Southwest Symposium on Mixed-Signal Design (SSMSD)* (2000), pp. 131–134
19. M.V. Heijningen, J. Compiet, P. Wambacq, S. Donnay, M.G.E. Engels, I. Bolsens, Analysis and experimental verification of digital substrate noise generation for Epi-type substrates. in *IEEE J. Solid St. Circ.* **35**(7), 1002–1008 (July 2000)
20. J.M. Cohn, D.J. Garrod, R.A. Rutenbar, L.R. Carley, KOAN/ANAGRAM II: new tools for device-level analog placement and routing. in *IEEE J. Solid St. Circ.* **26**(3), 330–342 (March 1991)
21. E. Malavasi, A. Sangiovanni-Vincentelli, Area routing for analog layout. in *IEEE T. Comput. Aid. D.* **12**(8), 1186–1197 (Aug 1993)
22. E. Malavasi, D. Pandini, Optimum CMOS stack generation with analog constraints. in *IEEE T. Comput. Aid. D.* **14**(1), 107–122 (Jan 1995)
23. J.D. Bruce, H.W. Li, M.J. Dallabetta, R.J. Baker, Analog layout using ALAS. in *IEEE J. Solid St. Circ.* **31**(2), 271–274 (Feb 1996)
24. D.A. Johns, K. Martin, *Analog Integrated Circuit Design* (Wiley, New York, 1997)
25. R.J. Baker, H.W. Li, D.E. Boyce, *CMOS Circuit Design, Layout, and Simulation* (IEEE Press, New York, 1997)
26. B. Razavi, *Design of Analog CMOS Integrated Circuits* (McGraw-Hill, New York, 2001)

Chapter 6
Experimental Results

6.1 Introduction

The proposed time-interleaved ADC [1] has been fabricated in a 0.18 μm one-poly six-metal CMOS process. The chip samples are packaged in 68-pin Ceramic Quad Flat-Pack (CQFP) packages. The successful measurement of a several-hundred MHz ADC is not a trivial task, and a special attention is necessary in the Printed-Circuit Board (PCB) design as well as the measurement setup that must ensure signal integrity. This chapter will present the design of the PCB with the consideration of high-frequency performance, and the testing setup in order to validate the design. The measurement results of the ADC will be exposed and a comparison with state-of-the-art low-voltage ADCs is presented.

6.2 The Prototype PCB Design

Unlike the low-frequency designs, the success of the PCB design for a high-speed mixed-signal system relies heavily on proper grounding. Only a large area solid ground plane is effective as compared with the single-point or multi-point grounding [2] alternative. Four-layer PCB is thus adopted in this design for better high-frequency performance, with ground planes as middle layers. The simplified block diagram of the PCB is shown in Fig. 6.1 with the top-view and bottom-view of the PCB shown in Fig. 6.2. Table 6.1 also demonstrates the equipments used in the measurement setup. The details on the PCB design will be explained in the following sub-sections.

6.2.1 The Floor Plan

The whole PCB is divided into three domains, namely Analog (AVDD), on-chip Digital (DVDD), as well as Board-level Digital (BVDD) domains, as shown in

S.-W. Sin et al., *Generalized Low-Voltage Circuit Techniques for Very High-Speed Time-Interleaved Analog-to-Digital Converters*, Analog Circuits and Signal Processing, DOI 10.1007/978-90-481-9710-1_6, © Springer Science+Business Media B.V. 2010

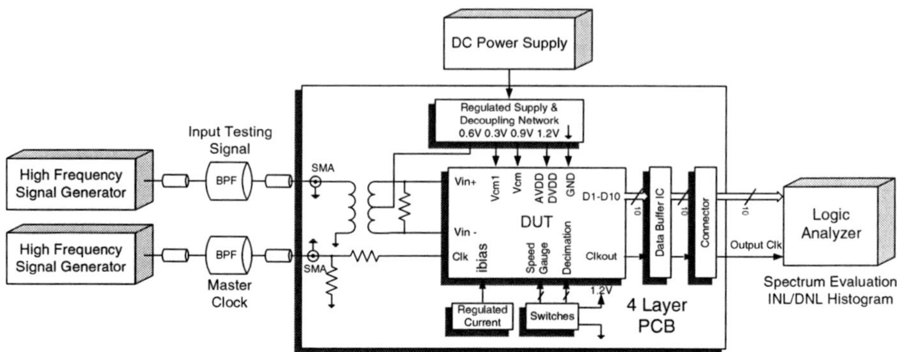

Fig. 6.1 PCB board diagram and experimental setup

Fig. 6.2, for better Electromagnetic-Compatibility (EMC). The PCB is designed such that no high-frequency signals cross the boundary between any two domains to avoid supply-noise coupling to AVDD from other domains, especially the BVDD domain which is very noisy since it contains the 10 b high-frequency digital output codes from the Device-Under-Test (DUT) and the data buffer IC.

For better high-frequency performance, Surface-Mount Device (SMD) is used throughout the whole PCB. High-speed devices are placed closest to the DUT (e.g. clock routing and its matching network, input network, and the data buffer IC) for shortest high-speed trace routing, while the regulated voltages and currents, as well as the control switches are placed at the outmost of the PCB.

6.2.2 Power Supply Considerations

The most effective grounding scheme in high-speed PCB designs is to use ground and supply planes to effectively reduce the resistance, and most importantly the inductance, within the power supply network by imaging the current-return paths just benefit the corresponding signal traces [3–7]. The whole internal second layer is used as a ground plane, while the internal third layer is used as both power and ground planes. In additions, ground planes are filled with the unused space in the signal-routing layers (first and fourth) for better EMC performance.

The power supplies are stabilized on-board by LC passive filtering composed by RF chokes and decoupling capacitors which are close to the DUT and the data buffer. To alleviate the self-resonant due to the Equivalent Series Inductance (ESL), the decoupling capacitors are composed by parallel combination of tantalum 220 μF, as well as ceramic 10 μF, 100 nF, 1 nF and 100 pF, to

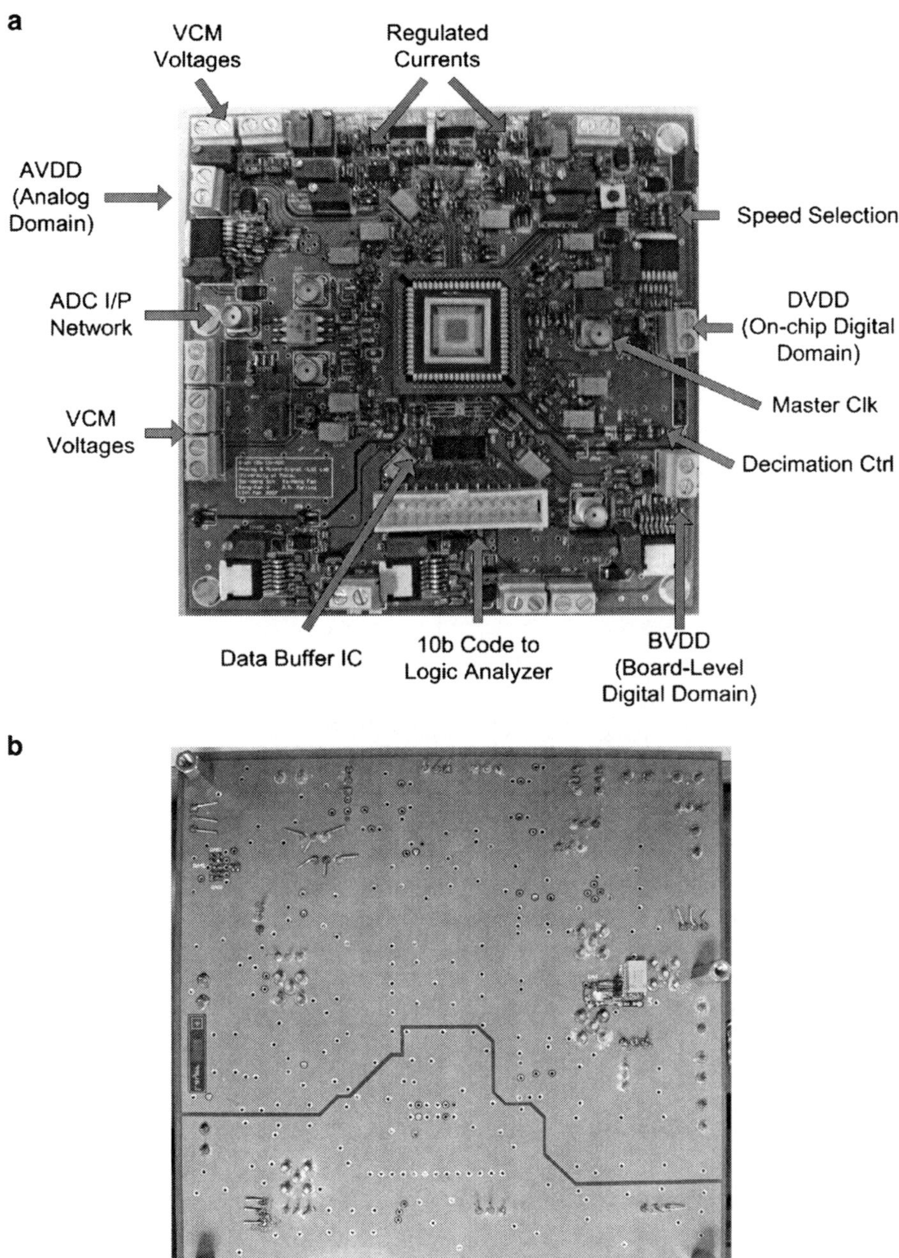

Fig. 6.2 (a) Top-view (b) Bottom-view of the four-layer PCB

achieve low power-bus impedance across a wide frequency range [3–12]. All the decoupling capacitors have nearest vias to the ground and power planes underneath.

6.2.3 Signal Trace Routing

The single-ended testing input signal from the SMA connector is converted to fully-differential before entering the DUT by using a Mini-Circuit RF wideband transformer Mini-Circuits T1-6T-KK81 + (0.015 − 300 MHz). The common-mode level of the fully-differential input signal is set by the center-tap of the secondary-side of the transformer. 50-Ω terminations are used to properly match the impedance of the SMA connectors, transformer, input and output impedance of the equipments, as well as the characteristic impedance of the trace. Small series damping resistors are also added at the input signal and clock routing to the chip, as well as the 10-b data output from the ADC to alleviate the resonant from the bondwire and the trace inductance. All the high-speed signal are routed on the first layer only, but not fourth layer, to avoid the noise coupling due to the extra vias inductance crossing the ground and power planes. Only the common-mode signal is routed through the fourth layer with its associate decoupling networks.

To ensure signal integrity throughout the routing of the PCB, all high-speed traces should be treated as transmission lines. The signal traces are routed on the ground plane which is a microstrip configuration, and its characteristic impedance depends on the trace width, distance to ground plane as well as the dielectric material. According to the foundry data, 12-mil trace-width (calculated with the help of PCB design software) are used throughout all high-speed signal trace to maintain constantly 50-Ω characteristic impedance. Round corners are used instead of 90° or 45° corners to reduce signal reflection at the trace corners. The PCB foundry also provides an impedance-control option during PCB fabrications to achieve more accurate control in the trace impedance.

6.3 Measurement Setup and Results

The prototype is characterized experimentally at various supply voltage, sampling rates under all speed options. The 10–b data from the ADC are captured with the Logic Analyzer (as shown in Fig. 6.1) and subsequently sent to the computer for further processing of the ADC output codes, including the time-domain decimal-equivalent of the 10–b digital codes, static performance evaluation like Differential and Integral-Nonlinearity (DNL & INL), as well as the evaluation of dynamic performance. As presented in Chapter 5, the large layout distance among different channels will induce gain-mismatch which can be easily calibrated through DSP post-processing. For testing purposes the gain-mismatches are calibrated using digital post-processing (by measuring and normalizing the rms gain from various channels) [13] after the output codes are capture by Logic Analyzer. Figure 6.3 demonstrates the overall equipments setup of the measurement, and Table 6.1 shows the equipment list. The performance of the ADC for all speed options are summarized in Table 6.2.

a

b

Fig. 6.3 (a) Experimental Setup and (b) Software evaluation platform of the prototype

Table 6.1 Testing equipment list

Agilent E3631A DC power supply (0–6 V/5 A, 0- $\pm$ 25 V/1 A)	Agilent Infiniium 54832D oscilloscope (1 GHz, 4 GSa/s)
Agilent E4424B ESG-AP Signal Generator (250 kHz – 2 GHz)	Agilent E5382A 17 channels single-ended flying leads
Agilent E4436B ESG-DP Signal Generator (250 kHz – 3 GHz)	Agilent Probes – 1161A Passive Probes, 1159A Differential Probes
Agilent 33250A 80 MHz function/Arbitrary waveform generator	Fluke 179 True RMS Multimeter
Agilent 16702B logic analyzer system w/16760A Module (800 MHz, 34 channels)	Mini-circuits SMA Passive Filters – BBP10.7, SLP1.9 – 550, SHP25 – 300
Rohde & Schwarz FSU8 Spectrum analyzer (20 Hz – 8 GHz)	Mini-circuits T1-6T-KK81+ (0.015 MHz –300 MHz) Transformer

Table 6.2 Performance summary of the ADC

Technology	0.18 μm CMOS (Vthn/Vthp $= 0.49/-0.48$ V)				
Nominal core voltage	1.8 V				
Resolution	10 b				
ADC core area	13.2 mm^2 (2.2 mm^2/channel)				
Supply voltage	1.2 V				
Full-scale input range	1.2 Vpp differential				
No. of channels	1	2	3	4	6
Sampling rate (MS/s)	60	120	180	240	360
ADC core power (mW)	85.2	144	232	287	426
DNL (LSB)	$+0.7/-0.7$	$+0.9/-0.7$	$+0.8/-0.7$	$+0.7/-0.6$	$+0.8/-0.6$
INL (LSB)	$+1.2/-1$	$+0.9/-1$	$+0.8/-1$	$+1/-1$	$+0.7/-1$
SNDR @ fin $= 25.1$ MHz	55 dB	54.8 dB	54 dB	54.5 dB	54 dB
ENOB @ fin $= 25.1$ MHz	8.9	8.8	8.7	8.8	8.7
SFDR @ fin $= 25.1$ MHz	72 dB	69 dB	65 dB	68 dB	65 dB
THD @ fin $= 25.1$ MHz	-68 dB	-66 dB	-67 dB	-67 dB	-60 dB

6.3.1 Time-Domain Digital Output Code Pattern

For functional verification, the time-domain digital output code pattern is obtained from the 10-b ADC output codes and is presented in Fig. 6.4, with all six channels enabled (i.e. $f_s = 360$ MHz with 60 MHz per channel, amplitude of $=-0.5$ dBFS and $f_{in} = 25$ MHz), which clearly shows that the output codes resemble a sine-wave pattern. For measurement purpose, the output codes are decimated by a factor of 5 on-chip to reduce the supply noise due to the bondwires' inductance.

6.3.2 Static Performance

The static performance of the ADC was characterized by traditional sine-wave histogram method [14] instead of ramp histogram, since accurate ramp testing signal is difficult to generate than a simple single-tone sinusoidal signal. Figure 6.5

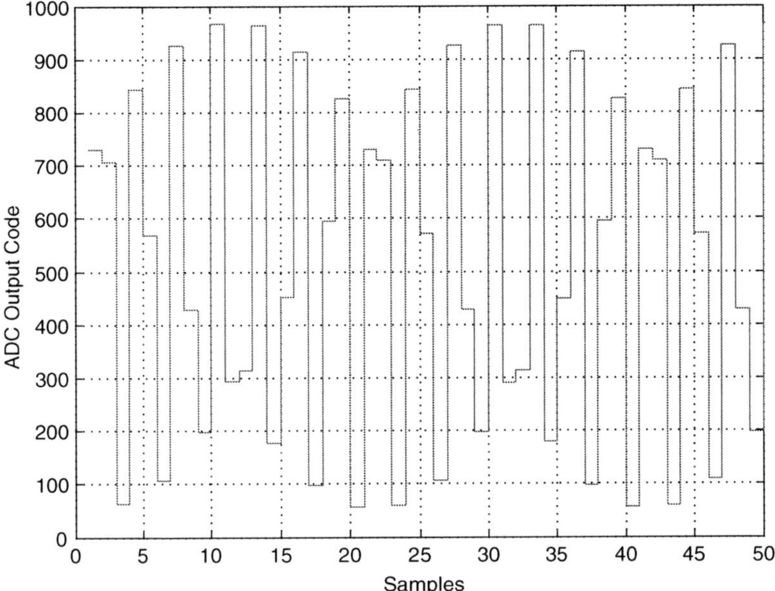

Fig. 6.4 Time-Domain digital output code of the ADC ($f_{in} = 25.2$ MHz, $f_s = 360$ MHz, output decimation by 5)

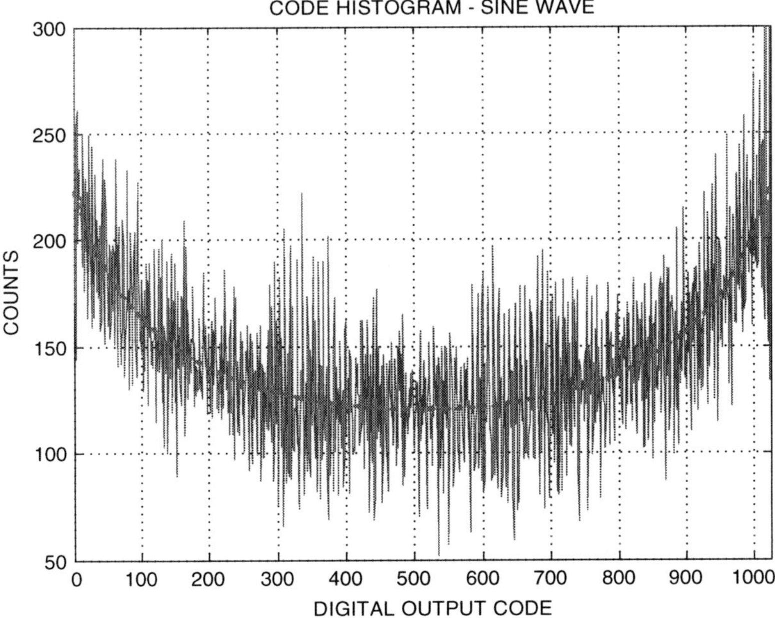

Fig. 6.5 Sine-wave histogram of the ADC for static performance evaluation (six channels, no. of samples = 220 k)

demonstrates the measured sine-wave histogram of the prototype, and after normalization process the DNL and the INL of the ADC can be calculated as shown in Fig. 6.6 (only six-channel option are shown here for simplicity). The measured DNL and INL are within 0.9/1.2 LSB for all speed options, as presented in Table 6.2, since time-interleaving only produces mismatch-type non-idealities that don't affect ADC's static linearity.

6.3.3 *Dynamic Performance*

The dynamic performance of the prototype ADC was also characterized to evaluate its performance at high speed.

Figure 6.7 shows the FFT output spectrum of the ADC with sampling frequency $f_s = 360$ MS/s (six channels with a decimation factor of 5) and input frequency $f_{in} = 25.2$ MHz, with and without gain-mismatch calibration. With calibration the SNDR improves from 43 to 54 dB. The gain-mismatch tones are located at 60 ± 25.2, 120 ± 25.2 and $180 - 25.2$ MHz, which correspond to 34.8, 85.2, 94.8, 145.2 and 154.8 MHz, respectively. These tones are mapped (and aliased) into the frequencies of 34.8, 13.2, 22.8, 1.2 and 10.8 MHz, respectively after decimation.

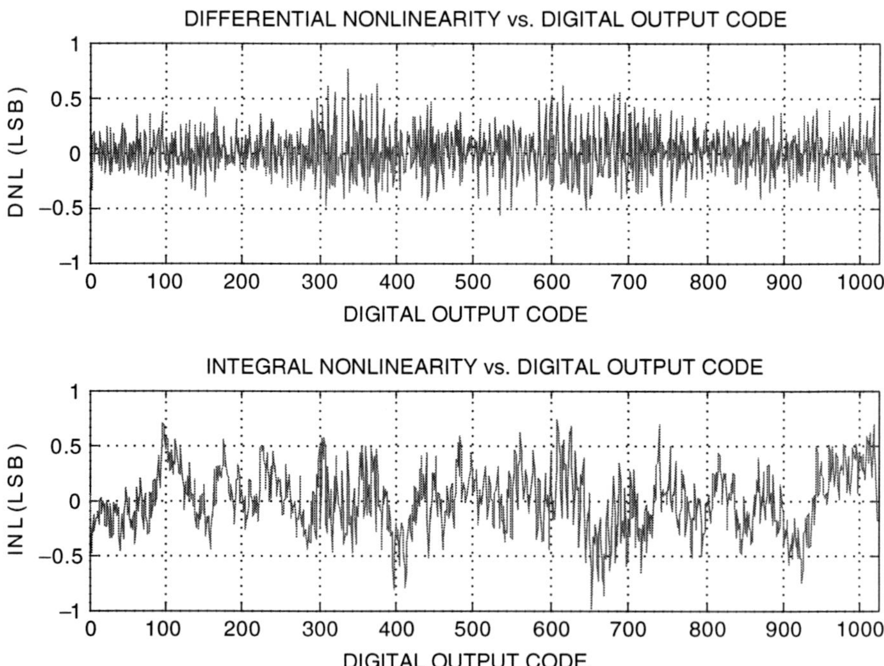

Fig. 6.6 Measured DNL and INL of the ADC (six channels)

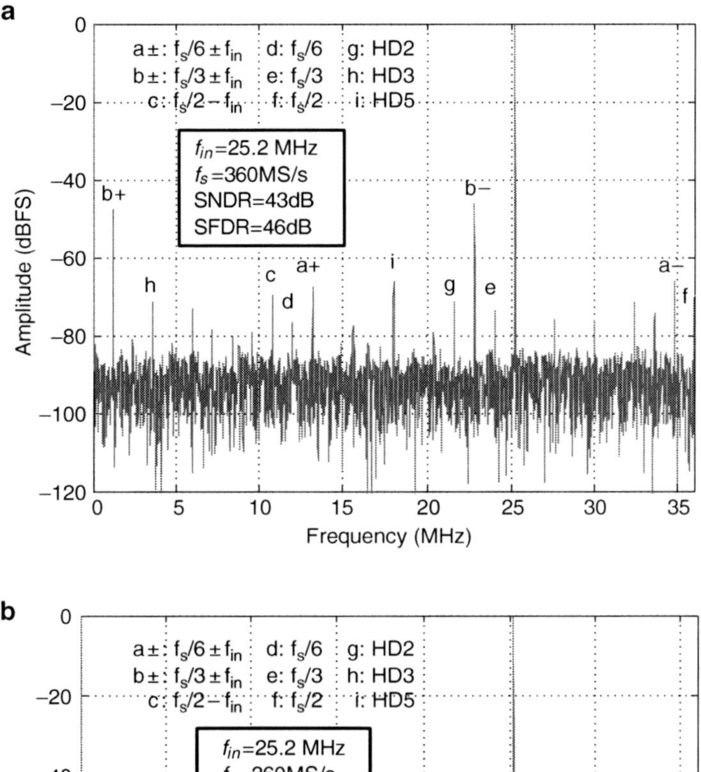

Fig. 6.7 Output spectrum (@*fs* = 360 MS/s) (**a**) without and (**b**) with gain mismatch calibration

Actually, from the ADC's layout (Fig. 5.9) the channel pairs (1, 4), (2, 5) and (3, 6) are arranged in close proximity to simplify the routing of the clock signals, and as a result the mismatch within a pair is minimized Consequently, the output spectrum equivalently contains a mismatch pattern that repeats with a period of $3/f_s$, as can be retained from Fig. 6.7a where the most prominent modulation sidebands appear at the frequencies of 120 ± 25.2 MHz before calibration, and the amplitude of these sidebands does not varies as the input frequency increases, an evidence of channel gain-mismatch rather than timing- or bandwidth-mismatch error. A simple mathematical deduction that confirms this result is provided in the Appendix D. The offset cancellation mechanisms described in Chapter 3 are also verified by the low offset induced tones at 12, 24 and 36 MHz (60, 120 and 180 MHz before decimation), as illustrated by Fig. 6.7b. Finally, the measured input referred noise of the ADC is $580 \ \mu V_{rms}$.

The ADC was also tested in all other speed options in various signal and sampling frequencies. Figure 6.8 shows a plot of SNDR versus input frequency at $f_s = 60$ MS/s per channel, which shows that the ADC has an Effective Resolution Bandwidth (ERBW) greater than 66 MHz. The bandwidth is mainly limited by the passive sampling network composed by $R = 1.6$ kΩ (the large resistance is used to achieve low-voltage cross-coupled sampling operation and the demultiplexing function) and the sampling capacitor $C_s = 1$ pF (without considering layout parasitics). Timing-skew induced SNDR reduction also starts to dominate at such high input frequency. Figure 6.9 shows another plot of SNDR versus sampling frequency with $f_{in} = 25.1$ MHz. The ADC maintains an SNDR greater than 54 dB (with 8.7 Effective-Number-Of-Bit – ENOB) for all speed options up to 60 MS/s per channel. At higher sampling frequencies the settling errors in the MDACs limit the ADC's performance. Figure 6.10 illustrates another plot of SNDR versus input signal amplitude with a peak SNDR achieved at $-0.5 \sim -0.1$ dBFS.

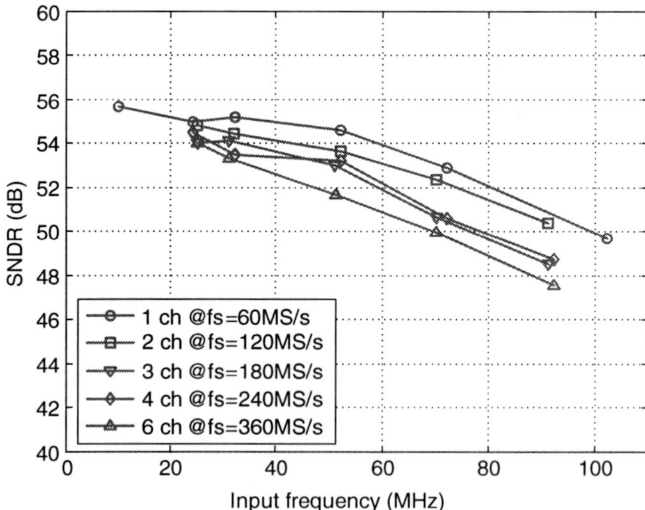

Fig. 6.8 A plot of SNDR versus f_{in} for all speed options ($f_s = 60$ MS/s per channel)

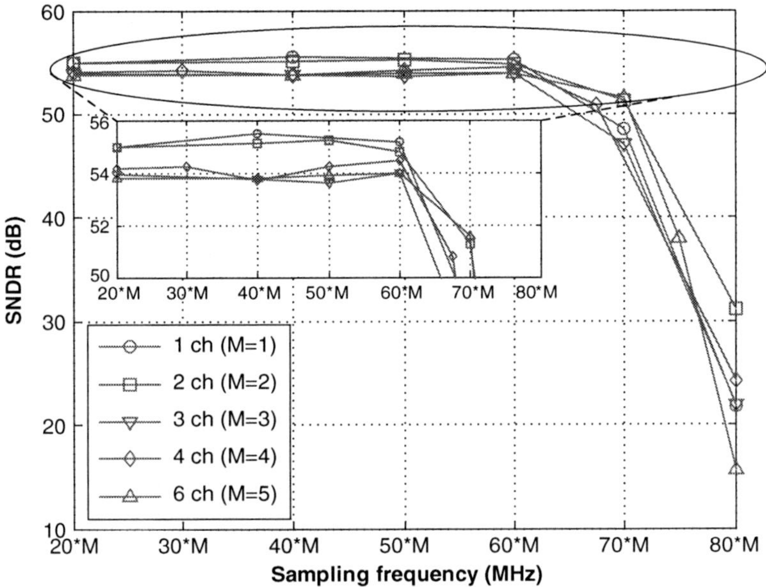

Fig. 6.9 A plot of SNDR versus sampling frequency per channel for all speed options (f_{in} = 25.2 MHz)

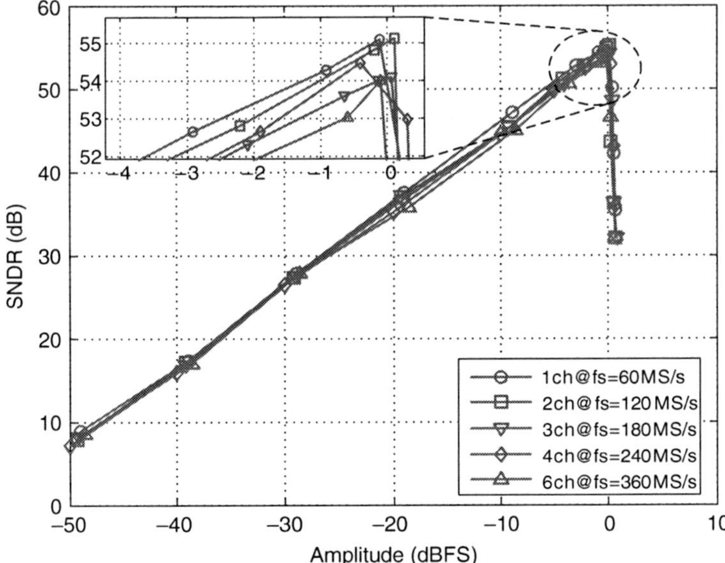

Fig. 6.10 A plot of SNDR versus input amplitude for all speed options (f_s = 60 MS/s per channel and f_{in} = 25.1 MHz)

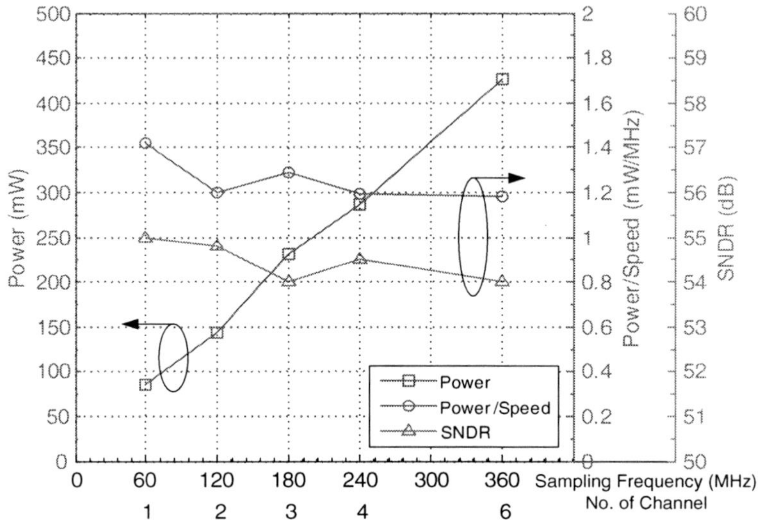

Fig. 6.11 A plot of power, power/speed and SNDR versus No. of channel for all speed options (f_s = 60 MS/s per channel and f_{in} = 25.1 MS/s)

To verify the power/speed scalability of the design, Fig. 6.11 presents a plot of power, power/speed ratio, as well as SNDR versus no. of channels and sampling frequencies. It shows that the power consumption is linearly scaled with the no. of active channels and the power/speed ratio remains approximately constant. The ADC consumes 85.2 mW @ 60 MS/s and 426 mW @ 360 MS/s.

6.4 Summary

This chapter has presented the setup as well as the results of the measurement of the low-voltage time-interleaved ADC that had utilized all the techniques presented in previous chapters, and the performance is summarized in Table 6.2. On the other hand, Table 6.3 exhibits the performance of the ADC and a benchmark of this work with state-of-the-art eight to 12 bit designs, within the same technology node. Usually, a Figure-Of-Merit (FOM) used to evaluate ADC's performance, designated here as FOM1, is given by [23]:

$$FOM1 = \frac{Power}{2^{ENOB} \cdot f_s}$$ (6.1)

However, since the supply voltage has been lowered the challenge in the optimization of ADC's performance is vastly increased because of: (i) the limited

Table 6.3 Performance benchmark with state-of-the-art designs

	This Work	JSSC05 [15]	JSSC07 [16]	JSSC04 [17]	ISSCC04 [18]	JSSC04 [19]	ISSCC05 [20]	ISSCC05 [21]	CICC06 [22]
Technology	**0.18 mm CMOS (nominal supply = 1.8 V)**								
Resolution	10 b	12 b	8 b	10 b	8 b	8 b	10 b	8 b	10 b
Supply Voltage	*1.2 V*	0.9 V	1 V	1.8 V	1.8 V	1.8 V	1.8 V	1.8 V	1.8 V
Supply Voltage (% of Nominal)	*67%*	50%	56%	100%	100%	100%	100%	100%	100%
Full-Scale Input Range	*1.2 Vpp*	0.5 Vpp	0.5 Vpp	1 Vpp	1.6 Vpp	0.8 Vpp	1.6 Vpp	1 Vpp	1.6 Vpp
Power/Speed Scalability	*Yes*	No	No	No	No	No	Yes	No	No
Sampling Rate	*360 MS/s*	5 MS/s	100 MS/s	100 MS/s	150 MS/s	1.6 GS/s	50 MS/s	200 MS/s	100 MS/s
ADC Core Power	426 mW	12 mW	30 mW	67 mW	71 mW	774 mW	35 mW	30 mW	76 mW
SNDR	54 dB	50 dB	41.5 dB	54 dB	46.8 dB	47.5 dB	55 dB	48 dB	57 dB
ENOB	8.7	8	6.6	8.7	7.5	7.6	8.8	7.7	9.2
FOM1 (Power/ $[2^{ENOB}*fs]$ pJ/converslon step)	2.8	9.4	3.1	1.6	2.6	2.5	1.6	0.7	1.3
FOM2 (Power* VDD/ $[2^{ENOB}*fs]$ pJ-V/ conversion step)	3.4	8.4	3.1	2.9	4.7	4.5	2.8	1.3	2.3

overdrive voltages of opamps and switches; (ii) the impossibility of applying many special circuit techniques due to the absence of floating switches; (iii) the fact that thermal noise doesn't decrease with the lowering of the voltage supply. Then, a more appropriate performance evaluator (FOM2) that will reflect better the degraded performance caused by reduced supply voltage can be used here and it is defined by [24, 25]:

$$FOM2 = \frac{Power}{2^{ENOB} \cdot f_s} V_{DD} \tag{6.2}$$

In this definition, the implemented ADC achieves a FOM2 of 3.4 pJ·V/step, which is comparable with several 1.8-V designs (i.e. with nominal supply) from Table 6.3. Even considering FOM2 the low-voltage ADCs (e.g. [15, 16] from Table 6.3) can't achieve the best performance when compared with nominal-supply designs, indicating a vastly large reduction in different performance characteristics basically inherited from the lower voltage. As shown in a comparison chart of Power per conversion step versus Speed in Fig. 6.12, the proposed ADC can successfully achieve very high speed with satisfactory performance under the stringent penalty of low-voltage environment, as compared with other low-voltage designs.

Table 6.3 also highlights key features of the proposed ADC, namely: (i) power/ speed scalability; (ii) only 67% usage of nominal supply voltage (indicating large headroom for future CMOS technology scaling); (iii) advantages in full-scale input range; (iv) very high-speed of operation with medium resolution.

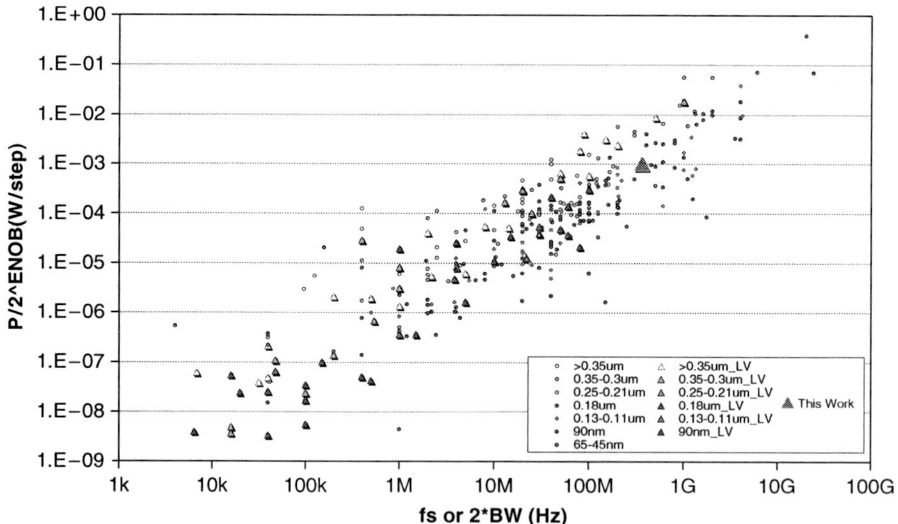

Fig. 6.12 Brief comparison of the state-of-art high-performance ADCs – Power per conversion step (Power/2^ENOB) versus Speed

References

1. S.-W. Sin, U. Seng-Pan, R.P. Martins, A 1.2-V 10-bit 60–360MS/s time-interleaved pipelined ADC in 0.18um CMOS with minimized supply headroom. IET Proc. Circuit. Device. Syst. **4**(1), 1–13 (Jan 2010)
2. H.W. Ott, *Noise Reduction Techniques in Electronic Systems* (Wiley, New York, 1998)
3. M.K. Armstrong, PCB design techniques for lowest-cost EMC compliance: Part 1. Electron. Comm. Eng. J. **11**, 185–194 (Aug 1999)
4. M.K. Armstrong, PCB design techniques for lowest-cost EMC compliance: Part 1. Electron. Comm. Eng. J., pp. 218–226 (Oct 1999)
5. M. Montrose, *EMC and the Printed Circuit Board: Design, Theory, and Layout Made Simple* (IEEE Press, New York, 1999)
6. K. Fowler, Grounding and shielding, Part 1 – Noise. in *IEEE Instru. Meas. Mag.*, pp. 41–44 (June 2000)
7. K. Fowler, Grounding and shielding, Part 2 – Grounding and return. in *IEEE Instru. Meas. Mag.*, pp. 45–48 (June 2000)
8. Texas Instrument, *The Bypass Capacitor in High-Speed Environments*, Application note (Nov 1996)
9. T.H. Hubing, J.L. Drewniak, T.P. Van Doren, D.M. Hockanson, Power bus decoupling on multilayer printed circuit boards. in *IEEE T. Electromagn. C.* **37**, 155–166 (May 1995)
10. J. Chen, M. Xu, T.H. Hubing, J.L. Drewniak, T.P. Van Doren, R.E. DuBroff, Experimental evaluation of power bus decoupling on a 4-layer printed circuit board, in *Proceedings of International Symposium on Electromagnetic Compatibility*, vol. 1 (2000)
11. J.K. Im, B. Choi, H. Kim, W. Ryu, Y.H. Yun, S.H. Ham, S.H. Kim, Y.H. Lee, J.H. Kim, Separated role of on-chip and on-PCB decoupling capacitors for reduction of radiated emission on printed circuit board. Proc. Int. Symp. Electromagn. C. **1**, 531–536 (2001)
12. U. Seng-Pan, R.P. Martins, J.E. Franca, *Design of Multirate Switched-Capacitor Circuits for Very High-Frequency Analog Front-End Filtering* (Springer, New York, 2005)
13. S.K. Gupta, M.A. Inerfield, J. Wang, A 1-GS/s 11-bit ADC with 55-dB SNDR, 250-mW power realized by a high bandwidth scalable time-interleaved architecture. in *IEEE J. Solid St. Circ.* **41**(12), 2650–2657 (Dec 2006)
14. J. Doernberg, H. Lee, D. Hodges, Full-speed testing of A/D converters. in *IEEE J Solid St. Circ.* **19**, 820–827 (Dec 1984)
15. J. Li, G. Ahn, D. Chang, U. Moon, A 0.9-V 12-mW 5-MSPS algorithmic ADC with 77-dB SFDR. in *IEEE J. Solid St. Circ.* **40**(4), 960–969 (April 2005)
16. P.Y. Wu, V.S.L. Cheung, H.C. Luong, in *IEEE J. Solid St. Circ.* **42**(4), 730–738 (April 2007)
17. J. Li, U. Moon, A 1.8-V 67-mW 10-bit 100-MS/s pipelined ADC using time-shifted CDS technique. in *IEEE J. Solid St. Circ.* **39**(9), 1468–1476 (Sept 2004)
18. S. Limotyrakis et al., A 150MS/s 8b 71mW time-interleaved ADC in 0.18μm CMOS, in *Digest of Technical Papers of IEEE International Solid-State Circuits Conference (ISSCC)* (Feb 2004), pp. 258–259
19. R. Taft, C. Menkus, M.R. Tursi, O. Hidri, V. Pons, A 1.8V 1.6GS/s 8b self-calibrating folding ADC with 7.26 ENOB at Nyquist frequency. in *IEEE J. Solid St. Circ.* **39**(12), 2107–2115 (Dec 2004)
20. I. Ahmed, D. Johns, A 50 MS/s (35 mW) to 1 kS/s (15 μW) power scalable 10b pipelined ADC with minimal bias current variation, in *Digest of Technical Papers of IEEE International Solid-State Circuits Conference (ISSCC)* (Feb 2005), pp. 280–281, 598
21. H.C. Kim, D.K. Jeong, W. Kim, A 30mW 8b 200MS/s pipelined CMOS ADC using a switched-opamp technique, in *Digest of Technical Papers of IEEE International Solid-State Circuits Conference (ISSCC)* (Feb 2005), pp. 284–285

22. K.E. Sankary, M. Sawan, 10-b 100-MS/s two-channel time-interleaved pipelined ADC, in *Proceedings of 2006 Custom Integrated Circuits Conference (CICC)* (Sept 2006), pp. 217–220
23. J. Goes, J.C. Vital, J.E. Franca, *Systematic Design for Optimization of Pipelined ADCs* (Kluwer, Boston, MA, 2001)
24. Y. Chiu, P.R. Gray, B. Nikolic, A 14-b 12-MS/s CMOS pipeline ADC with over 100-dB SFDR. in *IEEE J. Solid St. Circ.* **39**(12), 2139–2151 (Dec 2004)
25. K. Honda, M. Furuta, S. Kawahito, in *IEEE J. Solid St. Circ.* **42**(4), 757–765 (April 2007)

Chapter 7
Conclusions and Prospective for Future Work

7.1 Conclusions

The research work presented in this book led to the development of various techniques for the design of high-speed analog-to-digital converters under low-voltage environment imposed by CMOS technology scaling. The effectiveness of all the proposed techniques has been verified with a real IC prototype realization of a reconfigurable time-interleaved pipelined ADC. The novel circuit techniques utilized in this ADC design allow, simultaneously, the stretching of speed limits in the presence of low-voltage and a deep-submicron technology environment.

In Chapter 2 an overview of the impacts imposed by CMOS technology scaling has been presented. The performance degradation can be classified into two categories: device level and circuit level. Short-channel transistors generally exhibit poor characteristics over their long-channel counterparts, leading to degraded transconductance, output resistance, and limiting the intrinsic speed benefits brought from smaller feature sizes. Technology scaling also imposed a low-voltage design environment, reducing the design headroom in various circuit blocks, such as switches, current sources, opamps, and the unavailability of floating switches even forces the traditional SC circuits to fail. Existing solutions have been presented to overcome the absence of floating switches, including switches positioning, clock boosting, bootstrapped switches, switched- and reset-opamps, as well as switched-RC techniques. All of their pros and cons have been analyzed in detail, and although they could be utilized to resolve the fundamental drawbacks of floating switches, the limit of low-voltage still posed critical challenges in designing high-speed SC circuits, e.g. the design of CMFB, front-end interfaces to continuous-time signal, level-shifting required in low-voltage designs, increased sensitivity to process- and mismatch-variations, as well as the reduction in the achievable gain of the opamp designs.

In Chapter 3 all of the above challenges were analyzed individually, together with the presentation of novel solutions, namely the Virtual-Ground Common-Mode Feedback (VG-CMFB) with Output Common-Mode Error Correction (O-CMEC),

S.-W. Sin et al., *Generalized Low-Voltage Circuit Techniques for Very High-Speed Time-Interleaved Analog-to-Digital Converters*, Analog Circuits and Signal Processing, DOI 10.1007/978-90-481-9710-1_7, © Springer Science+Business Media B.V. 2010

Crossed-Coupled Passive Sampling Interface (CCPSI), Voltage-Controlled Level-Shifting (VCLS), Low-Voltage Finite Gain-and-Offset Compensation (LV-FGC & LV-OC), as well as the Feedback Current Biasing (FBCB), to solve the various types of low-voltage limitations. The ultimate goals of these solutions are targeting the formation of a set of advanced circuit techniques that enable low-voltage SC designs and reduce the performance gap between these and traditional nominal-voltage SC designs.

In Chapter 4 the theories behind time-interleaved ADC were introduced. Time-interleaving can multiply the speed of the ADC above the technology limits, but it is susceptible to various types of channel mismatch errors, including offset, gain, timing and sampling bandwidth. The effects of all these mismatches will be mathematically analyzed and a closed-form SNDR formula was derived such that the performance of such systems can be evaluated systematically in very early stages of the design process.

Chapter 5 presented the implementation details of the IC prototype which was a 1.2 V 10 b 60–360 MS/s time-interleaved pipelined ADC employing all the techniques discussed before. The prototype was fabricated in a 0.18 μm CMOS technology ($V_{thn} = 0.63$ V and $V_{thp} = -0.65$ V for mid-supply floating switches). Simultaneous low-voltage and high-speed requirements create stringent design challenges and the whole design has been thoroughly investigated with verification from top-level evaluation during the design phase to transistor-level and post-layout simulation with consideration of process and mismatch variations.

Chapter 6 exhibited the experimental setup and verifications of the prototype ADC. Special attention was paid to the design of the PCB due to the nature of the high-speed and mixed-signal IC. The prototype has been thoroughly tested with different sampling rates as well as input signal frequencies and amplitudes, in both static and dynamic ways. The ADC achieved better than 0.9/1.2 LSB of DNL and INL, and greater than 54 dB SNDR for all speed options, with approximately linearly scaled power consumption from 85 mW @ 60 MS/s to 426 mW @ 360 MS/s, with die area of 13.2 mm^2 (2.2 mm^2/channel). The measurement results have shown that the ADC operates with exceptional consistency with the theoretical results, thus consolidating the effectiveness of all the design techniques presented previously.

7.2   Prospective for Future Work

This work presented in this book serves primarily as a silicon proof of the various proposed low-voltage circuit techniques that can be efficiently applied in the design of very high-speed ADCs. In a prospective for future work, further research paths are worth to be explored and investigated. The following aspects, in particular, are very interesting to pursue.

7.2.1 Low-Noise and Low-Voltage Circuit Techniques Implementation

Reset-opamp techniques proved to be efficient high-speed solutions to face the floating switches problem, but it suffers from larger opamp circuit's noise when compared to traditional SC circuit implementations due to the intensive usage of opamp's output. In addition, the low-voltage gain-compensation technique relies on the use of auxiliary opamps, which also add to the circuit noise. In fact, one of the dominant error sources in this prototype implementation is the thermal noise, thus low-noise low-voltage circuit techniques could be one of the key research areas to explore in high-speed high-resolution ADC designs.

7.2.2 Fully Dynamic ADC Implementations

Pipelined ADCs utilize opamps to achieve accurate residue amplification, which leads to larger power consumption due to the static power of the opamps. There are increasing trends to use dynamic elements to replace the static elements in the circuits, e.g. the opamps may be replaced by comparators [1], zero-crossing detectors [2] or even simply by inverters [3, 4], in order to greatly reduce the power consumption. However, the nonlinear characteristics of such dynamic elements will impose performance limitations in the ADC, thus presenting a new and challenging research direction related with the development of a new class of sampled-data systems that will use only dynamic circuits.

Also, some ADC architectures are intrinsically dynamic, like Successive Approximation Register (SAR) ADC [5–7] that uses only a comparator and a passive capacitor DAC array to achieve data conversion. The speed of a high-resolution SAR ADC is inherently low; however, due to the deep-submicron technology scaling the speed of all dynamic circuits used in the SAR can be enhanced. In addition, recent advances in design techniques, such as asynchronous processing [5], have enabled the design of high-speed power-efficient SAR ADC implementations, which can become one of the main streams of future ADC research.

7.2.3 SC Circuits Implemented with Open-Loop Amplifiers

Traditional pipelined ADCs also utilize opamps into the negative feedback loop to achieve accurate residue amplification. However, using opamps in negative feedback creates stability problems and to ensure stability the non-dominant pole must be located at higher frequency above the intended closed-loop bandwidth of the amplifiers. This non-dominant pole often limits the achievable speed of the opamps, but this would not bring any problem if the amplifier is used in an open-loop configuration [8], which is a new topic that has emerged recently. However, open-loop amplifiers are sensitive to process variation and nonlinearity, thus the implementation is challenging and still requires extensive research in the future.

7.2.4 Digital Calibration

Aggressive CMOS technology scaling permits robust designs of digital circuits and microprocessors which enable the implementation of complex algorithms, although imposing performance degradation into analog circuit designs. Digital Calibration [8], which relies on the robust processing power of digital circuits to calibrate analog nonidealities, has become a hot research topic recently and proven to be an efficient technique in deep-submicron data converter designs. However, dynamic errors such as settling errors, timing-skew and bandwidth mismatches, as well as nonlinearities, are difficult to be calibrated and usually the corresponding algorithms are quite complex and lead to large power consumption and substrate noise in the digital part of the data converters. As a result, this would also be one of the important research directions to be followed in the future where the development of efficient and compact digital calibration techniques to compensate these errors should be addressed.

References

1. J.K. Fiorenza, T. Sepke, P. Holloway, C.G. Sodini, H.-S. Lee, Comparator-based switched-capacitor circuits for scaled CMOS technologies. in *IEEE J. Solid-St. Circ.* **41**(12), 2658–2668 (Dec 2006)
2. L. Brooks, H.-S. Lee, A zero-crossing-based 8-bit 200 MS/s pipelined ADC. in *IEEE J. Solid-St. Circ.* **42**(12), 2677–2687 (Dec 2007)
3. J. Li, U.-K. Moon, A 1.8-V 67-mW 10-bit 100-MS/s pipelined ADC using time-shifted CDS technique. in *IEEE J. Solid-St. Circ.* **39**(9), 1468–1476 (Sept 2004)
4. Y. Chae, I. Lee, G. Han, A 0.7V 36μW 85dB-DR audio ΔΣ modulator using class-C inverter, in *Digest of Technical Papers of IEEE International Solid-State Circuits Conference (ISSCC)* (Feb 2008), pp. 490–491,630
5. M.S.W. Chen, R.W. Brodersen, A 6b 600 MS/s 5.3mW asynchronous ADC in 0.13μm CMOS, in *IEEE ISSCC Digest of Technical Papers* (Feb 2006), pp. 574–575
6. J. Craninckx, G. van der Plas, A 65fJ/conversion-step 0-to-50MS/s 0-to-0.7mW 9b charge-sharing SAR ADC in 90nm digital CMOS, in *IEEE ISSCC Digest of Technical Papers* (Feb 2007), pp. 246–247
7. V. Giannini, P. Nuzzo, V. Chironi, A. Baschirotto, G. Van der Plas, J. Craninckx, An 820μW 9b 40MS/s noise-tolerant dynamic-SAR ADC in 90nm digital CMOS, in *IEEE ISSCC Digest of Technical Papers* (Feb 2008), pp. 238–239
8. B. Murmann, B.E. Boser, A 12-bit 75-MS/s pipelined ADC using open-loop residue amplification. in *IEEE J. Solid-St. Circ.* **38**(12), 2040–2050 (Dec 2003)

Appendix A
Operating Principle of VG-CMFB
with O-CMEC

This appendix provides a detailed mathematical analysis of the operating principle associated with VG-CMFB + O-CMEC techniques, presented before in Section 3.2.4, as well as illustrates the derivation of the equation that is necessary for determining the capacitor ratio selection.

The derivation is based on the VG-CMFB and O-CMEC structures already presented in Fig. 3.5 plus the external passive network configuration from Fig. 3.3. Referring to Fig. 3.5, V_{bias} and $V_{tail,ref}$ are the nominal values (or zero-CM-error operating point) of the node V_{CMFB} and V_{tail}, respectively, and $V_{G,ref}$ is the desired value of the virtual ground CM voltage $V_{G,CM}$. Assuming A_{CM} is the CM gain from the node V_{CMFB} to the output CM level $V_{out,CM}$, then the following relationship holds for a negative CMFB loop:

$$V_{out,CM} = -A_{CM}(V_{CMFB} - V_{bias}) \tag{A1}$$

or

$$V_{CMFB} = -\frac{V_{out,CM}}{A_{CM}} + V_{bias} \approx V_{bias} \tag{A2}$$

with the assumption that $A_{CM} \gg 1$. Considering that the circuit is in the steady-state the O-CMEC should inject no correction charges into the V_{CMFB} node in phase 2 if no output CM error occurs, thus $\Delta q = 0$ which yields:

$$2C_3(V_{out,CM}[\phi 2] - V_{out,CM}[\phi 1]) + C_4(V_1 - V_2) \\ - (2C_3 + C_4)(V_{CMFB}[\phi 2] - V_{CMFB}[\phi 1]) = 0 \tag{A3}$$

By substituting (A2) in Eq. (A3) it can be simplified to the following design equation:

$$V_{out,CM}[\phi 2] - V_{out,CM}[\phi 1] = \frac{C_4}{2C_3}(V_2 - V_1) \tag{A4}$$

which defines the relationship between the output CM level in two phases with the capacitor ratio and the reference voltages V_1 and V_2. Notice that by setting $\Delta q = 0$ in phase 1 the same Eq. (A4) can be derived, implying that the O-CMEC circuit will not inject charges in both phases if there is no CM error in the steady-state.

The derivation is not complete until the virtual ground potential is determined. Referring to Fig. 3.3 in phase 1 the opamp reset implies that

$$V_{tail}[\phi 1] = V_{G,CM}[\phi 1] - V_{GS1,CM} = V_{out,CM}[\phi 1] - V_{GS1} \tag{A5}$$

where V_{GS1} is the CM gate-source voltage of the differential-pair transistors M1A and M1B in Fig. 3.4, which remains constant as long as the tail current source does not vary. A charge conservation equation can be written at node V_{CMFB} in phase 1 and rearranged as:

$$\begin{aligned}
(C_2 + 2C_3 + C_4)&(V_{CMFB}[\phi 1] - V_{CMFB}[\phi 2]) + C_1 V_{CMFB}[\phi 1] \\
&- C_2(V_{tail}[\phi 1] - V_{tail}[\phi 2]) - C_1 V_{tail}[\phi 1] - C_1(V_{bias} - V_{tail,ref}) \\
&- 2C_3(V_{out,CM}[\phi 1] - V_{out,CM}[\phi 2]) + C_4(V_1 - V_2) = 0
\end{aligned} \tag{A6}$$

Similarly an equation can be written at node V_{CMFB} in phase 2:

$$\begin{aligned}
-(C_2 + 2C_3 + C_4)&(V_{CMFB}[\phi 1] - V_{CMFB}[\phi 2]) + C_2(V_{tail}[\phi 1] - V_{tail}[\phi 2]) \\
&+ 2C_3(V_{out,CM}[\phi 1] - V_{out,CM}[\phi 2]) - C_4(V_1 - V_2) = 0
\end{aligned} \tag{A7}$$

Summing (A6) and (A7) it yields

$$V_{CMFB}[\phi 1] - V_{tail}[\phi 1] = V_{bias} - V_{tail,ref} \tag{A8}$$

Substituting (A2) and (A8) with $V_{tail,ref} = V_{G,ref} - V_{GS1}$ into (A8) it will imply:

$$V_{G,CM}[\phi 1] \approx V_{G,ref} \tag{A9}$$

which indicates that the virtual ground CM voltage in phase 1 is stabilized to the reference voltage $V_{G,ref}$.

The virtual ground CM voltage in phase 2 can be evaluated by subtracting (A6) from (A7), with substitution of (A8) as:

$$\begin{aligned}
2(C_2 + 2C_3 + C_4)&(V_{CMFB}[\phi 1] - V_{CMFB}[\phi 2]) - 2C_2(V_{tail}[\phi 1] - V_{tail}[\phi 2]) \\
&- 4C_3(V_{out,CM}[\phi 1] - V_{out,CM}[\phi 2]) - 2C_4(V_1 - V_2) = 0
\end{aligned} \tag{A10}$$

Also the following equations will hold:

$$V_{CMFB}[\phi 1] \approx V_{CMFB}[\phi 2] \approx V_{bias} \tag{A11}$$

$$V_{tail}[\phi 1] - V_{tail}[\phi 2] = V_{G,CM}[\phi 1] - V_{G,CM}[\phi 2] \tag{A12}$$

Finally by substituting (A9), (A11), (A12) and the design Eq. (A4) into (A10) it yields:

$$V_{G,CM}[\phi 2] \approx V_{G,ref} \tag{A13}$$

which indicates that the virtual ground CM voltage in phase 2 is also stabilized to $V_{G,ref}$.

Appendix B
Mathematical Analysis of Bandwidth Mismatches

This appendix presents the detail derivations of the closed-form SNDR expression for time-interleaved ADC under bandwidth mismatch effect as discussed in Section 4.7. It would be necessary to recall the fact that for a sinusoidal input signal with frequency of $\omega_0 = 2\pi f_0$ the weights of modulation sidebands can be expressed as (4.38), which is repeated here:

$$A_k(\omega_0 + k\frac{2\pi}{MT}) = \frac{1}{M}\sum_{m=0}^{M-1} H_m(\omega_0)e^{-jkm\frac{2\pi}{M}} = \frac{a}{M}\sum_{m=0}^{M-1}\frac{e^{-jkm\frac{2\pi}{M}}}{1 + j\omega\tau(1 + \tau_m)} \tag{B1}$$

where for $k = 0$ $A_0(\omega_0)$ corresponds to the signal component, and for $k = 1$ to M-1 A_k correspond to M-1 different sideband components. Then, the expected values of the signal component $|A_0(\omega_o)|^2$ can be evaluated by substituting $k = 0$ into (B1) and multiplying it by its complex conjugate as follows:

$$E\left[|A_0(\omega_0)|^2\right] = \frac{a^2}{M^2}E\left[\left|\sum_{m=0}^{M-1}\frac{1}{1 + j\omega_0\tau(1 + \tau_m)}\right|^2\right] \approx \frac{a^2}{1 + \omega_0^2\tau^2} \tag{B2}$$

which shows that the signal component is approximately scaled by the RC-filter for small values of τ_m ($\tau_m \ll 1$). The expected value of the sideband components can also be calculated from (B1) leading to:

$$E\left[\left|A_k(\omega_0 + k\frac{2\pi}{MT})\right|^2\right]$$

$$= \frac{a^2}{M^2}E\left[\sum_{m=0}^{M-1}\sum_{n=0}^{M-1}\frac{e^{-jkm\frac{2\pi}{M}}}{1 + j\omega_0\tau(1 + r_m)} \cdot \frac{e^{jkn\frac{2\pi}{M}}}{1 - j\omega_0\tau(1 + r_n)}\right]$$

$$= \frac{a^2}{M^2}\sum_{m=0}^{M-1}\sum_{n=0}^{M-1}e^{-jk(m-n)\frac{2\pi}{M}}E\left[\frac{1}{1 + j\omega_0\tau(r_m - r_n) + \omega_0^2\tau^2(1 + r_m + r_n + r_m r_n)}\right]$$

$$\tag{B3}$$

where, in this case, the second order term $r_m r_n$ in (B3) cannot be neglected since in the following derivations it will be required to use the statistical values of $E[r_m] = 0$ and $E[r_m^2] = \sigma_r^2$ and, besides this, it will constitute the most significant contribution to the sideband components. However, in (B2) such approximation is valid because the amplitude of the signal component is, in principle, much larger than that of the sideband. The evaluation of the expected value of the sideband components in (B3) is quite complex and only key steps will be addressed here. The double summation in (B3) can be analyzed in two parts for $m = n$ (containing M terms) and $m \neq n$ (contain $M^2 - M$ terms) in the following way:

$$Part\ 1 = \frac{a^2}{M^2} \sum_{\substack{m=0 \\ m=n}}^{M-1} \sum_{n=0}^{M-1} E\left[\frac{1}{1 + \omega_0^2 \tau^2 (1 + 2r_m + r_m^2)}\right]$$

$$= \frac{a^2}{M^2} \sum_{\substack{m=0 \\ m=n}}^{M-1} \sum_{n=0}^{M-1} \frac{1}{1 + \omega_0^2 \tau^2} E\left[\frac{1}{1 + A}\right] \approx \frac{a^2}{M^2} \sum_{\substack{m=0 \\ m=n}}^{M-1} \sum_{n=0}^{M-1} \frac{1}{1 + \omega_0^2 \tau^2} E\left[1 - A + A^2\right]$$

$$\tag{B4}$$

$$Part\ 2 = \frac{a^2}{M^2} \sum_{\substack{m=0 \\ m\neq n}}^{M-1} \sum_{n=0}^{M-1} e^{-jk(m-n)\frac{2\pi}{M}} E\left[\frac{1}{1 + j\omega_0\tau(r_m - r_n) + \omega_0^2 \tau^2 (1 + r_m + r_n)}\right]$$

$$= \frac{a^2}{M^2} \sum_{\substack{m=0 \\ m\neq n}}^{M-1} \sum_{n=0}^{M-1} \frac{e^{-jk(m-n)\frac{2\pi}{M}}}{1 + \omega_0^2 \tau^2} E\left[\frac{1}{1 + B}\right] \approx \frac{a^2}{M^2} \sum_{\substack{m=0 \\ m\neq n}}^{M-1} \sum_{n=0}^{M-1} \frac{e^{-jk(m-n)\frac{2\pi}{M}}}{1 + \omega_0^2 \tau^2} E\left[1 - B + B^2\right]$$

$$\tag{B5}$$

where

$$A = \frac{\omega_0^2 \tau^2 (2r_m + r_m^2)}{1 + \omega_0^2 \tau^2}, \quad B = \frac{\omega_0^2 \tau^2 (r_m + r_n) + j\omega_0\tau(r_m - r_n)}{1 + \omega_0^2 \tau^2}$$

should be small for $r_m \ll 1$ leading to a valid Taylor Series expansion in (B4) and (B5). The expansion can also be evaluated up to the second order because A^3 and B^3 contain at least a third order term like r_m^3 which is negligible compared to the main contribution from r_m^2. The expected values can be calculated as:

$$E[A] = \frac{\omega_0^2 \tau^2}{1 + \omega_0^2 \tau^2} \left(2E[r_m] + E[r_m^2]\right) = \frac{\omega_0^2 \tau^2 \sigma_r^2}{1 + \omega_0^2 \tau^2} \tag{B6}$$

$$E[B] = \frac{\omega_0^2 \tau^2}{1 + \omega_0^2 \tau^2} \left(E[r_m] + E[r_n]\right) + \frac{j\omega_0\tau}{1 + \omega_0^2 \tau^2} \left(E[r_m] - E[r_n]\right) = 0 \tag{B7}$$

since r_m and r_n are random variables with zero means. Furthermore, for the second order term

$$E[A^2] = \left(\frac{\omega_0^2 \tau^2}{1 + \omega_0^2 \tau^2}\right)^2 \left(4E[r_m^2] + 4E[r_m^3] + E[r_m^4]\right) \approx \left(\frac{\omega_0^2 \tau^2}{1 + \omega_0^2 \tau^2}\right)^2 (4\sigma_r^2) \quad \text{(B8)}$$

$$E[B^2] = \left(\frac{\omega_0 \tau}{1 + \omega_0^2 \tau^2}\right)^2 E\left[\omega_0^2 \tau^2 (r_m + r_n)^2 + 2j\omega_0 \tau (r_m^2 - r_n^2) - (r_m - r_n)^2\right] \quad \text{(B9)}$$

Since r_m and r_n have zero mean and are independent (uncorrelated) (B9) can be evaluated considering $E[r_m^2] = E[r_n^2] = \sigma_r^2$ and $E[r_m r_n] = 0$ which will lead to:

$$E[B^2] = \left(\frac{\omega_0 \tau}{1 + \omega_0^2 \tau^2}\right)^2 \left[2\sigma_r^2 (\omega_0^2 \tau^2 - 1)\right] \quad \text{(B10)}$$

Then substituting (B6) and (B8) into (B4) it yields:

$$\begin{aligned}
Part\ 1 &= \frac{a^2}{M^2} \sum_{\substack{m=0 \\ m=n}}^{M-1} \sum_{n=0}^{M-1} \frac{1}{1 + \omega_0^2 \tau^2} \left[1 - \frac{\omega_0^2 \tau^2 \sigma_r^2}{1 + \omega_0^2 \tau^2} + \left(\frac{\omega_0^2 \tau^2}{1 + \omega_0^2 \tau^2}\right)^2 4\sigma_r^2\right] \\
&= \frac{a^2}{M} \frac{1}{1 + \omega_0^2 \tau^2} \left[1 + \left(\frac{\omega_0 \tau}{1 + \omega_0^2 \tau^2}\right)^2 \sigma_r^2 (3\omega_0^2 \tau^2 - 1)\right]
\end{aligned} \quad \text{(B11)}$$

and, similarly, by substituting (B7) and (B10) into (B5):

$$\begin{aligned}
Part\ 2 &= \frac{a^2}{M^2} \sum_{\substack{m=0 \\ m \neq n}}^{M-1} \sum_{n=0}^{M-1} \frac{e^{-jk(m-n)\frac{2\pi}{M}}}{1 + \omega_0^2 \tau^2} \left[1 + \left(\frac{\omega_0 \tau}{1 + \omega_0^2 \tau^2}\right)^2 \left[2\sigma_r^2 (\omega_0^2 \tau^2 - 1)\right]\right] \\
&= \frac{a^2}{M^2} \frac{1}{1 + \omega_0^2 \tau^2} \left[1 + \left(\frac{\omega_0 \tau}{1 + \omega_0^2 \tau^2}\right)^2 \left[2\sigma_r^2 (\omega_0^2 \tau^2 - 1)\right]\right] \sum_{\substack{m=0 \\ m \neq n}}^{M-1} \sum_{n=0}^{M-1} e^{-jk(m-n)\frac{2\pi}{M}}
\end{aligned}$$

$$\text{(B12)}$$

where

$$\sum_{\substack{m=0 \\ m \neq n}}^{M-1} \sum_{n=0}^{M-1} e^{-jk(m-n)\frac{2\pi}{M}} = \sum_{m=0}^{M-1} \sum_{n=0}^{M-1} e^{-jk(m-n)\frac{2\pi}{M}} - \sum_{\substack{m=0 \\ m=n}}^{M-1} \sum_{n=0}^{M-1} e^{-jk(m-n)\frac{2\pi}{M}} = 0 - M \quad \text{(B13)}$$

leading to the following simplification of part 2:

$$Part\ 2 = -\frac{a^2}{M}\frac{1}{1+\omega_0^2\tau^2}\left[1+\left(\frac{\omega_0\tau}{1+\omega_0^2\tau^2}\right)^2\left[2\sigma_r^2(\omega_0^2\tau^2-1)\right]\right] \qquad (B14)$$

Finally, the sideband components can be calculated by the addition of both parts 1 and 2:

$$E\left[\left|A_k(\omega_0+k\frac{2\pi}{MT})\right|^2\right] = \frac{a^2}{M}\frac{1}{1+\omega_0^2\tau^2}\left(\frac{\omega_0\tau}{1+\omega_0^2\tau^2}\right)^2\left[\sigma_r^2(\omega_0^2\tau^2+1)\right]$$

$$= \frac{a^2}{M}\frac{1}{1+\omega_0^2\tau^2}\frac{\omega_0^2\tau^2\sigma_r^2}{1+\omega_0^2\tau^2} \qquad (B15)$$

From (B15) it can be concluded that the expected values of the various sidebands are independent of the frequency index k, i.e. all the sidebands have the same expected distortion power. As a result, the total distortion power should be equal to (B15) multiplied by M-1, and the final SNDR can be calculated by substituting (B2) and (B15) into the following equation:

$$SNDR = 10\ \log_{10}\left[\frac{E[|A_0(\omega_o)|^2]}{E\left[\sum_k |A_k(\omega_o+k\frac{2\pi}{MT})|^2\right]}\right] dB \qquad (B16)$$

Then the SNDR of the time-interleaved ADC under bandwidth mismatch can be finally calculated as:

$$SNDR = 10\ \log_{10}\left[1+\omega_0^2\tau^2\right] - 20\ \log_{10}[\omega_0\tau\sigma_r] - 10\ \log_{10}\left[1-\frac{1}{M}\right] dB \qquad (B17)$$

Appendix C
Noise Analysis of Advanced Reset-Opamp Circuits

Cross-Coupled Front-End S/H

This section provides a thermal noise analysis of the front-end S/H with crossed-coupled passive sampling circuit, as presented in Fig. 5.3 and Table 3.1. In Fig. 5.3 the circuit noise is calculated as follows: In phase 1 (Fig. 5.7a), the thermal noise sampled in the capacitors $C_1 = C_3 = C$ will be transferred to the output in phase 2:

$$\overline{v^2_{n,kT/C}[\phi 1]} = 2kT/C \tag{C1}$$

and the noise sampled in capacitor C_3 is still equal to kT/C since the capacitor is reset by the switch in phase 1, independent of opamp finite bandwidth. For the capacitor $C_2 = C$, since the capacitor is connected to the virtual ground of the opamp, the noise generated by the switch resistance R_{S3} is no longer like a kT/C noise, but shaped by finite opamp bandwidth (assuming the bandwidth of all the circuits in this section are dominated by the opamp, and the bandwidth contribution from switches are neglected):

$$\overline{v^2_{n,Rs2}[\phi 1]} = 4kTR_{s2,A} \cdot \left| H_{C2,\phi 1}(0) \right|^2 \cdot \frac{1}{4} \omega_{GBW\_\phi 1} = kTR_{s2,A}\omega_{GBW\_\phi 1} \tag{C2}$$

where $H_{Cs2,\phi 1}(0) = 1$ is the low-frequency noise voltage transfer function from the noise source of switch $R_{s2,A}$ to the output, and $\omega_{GBW\_\phi 1}$ is the gain-bandwidth product of the opamp in phase 1. In addition, since the capacitor C_2 is connected between the virtual grounds of the opamp for offset compensation purpose, the capacitor equivalently samples twice the noise power of the opamp when compared to the normal configuration with the top plate of the capacitor connected to the virtual ground and the bottom plate connected to signal ground. The noise sampled to this capacitor is also transferred to the output in phase 2:

$$\overline{v^2_{n,opamp}[\phi 1]} = 2^* S_{opamp}(0) \cdot \left| H_{\phi 1}(0) \right|^2 \cdot \frac{1}{4} \omega_{GBW\_\phi 1} = \frac{1}{2} S_{opamp}(0)\omega_{GBW\_\phi 1} \tag{C3}$$

where $S_{opamp}(f)$ is the opamp's input referred noise power spectral density, which depends only on the opamp topology. Moreover, due to the resetting feature of RO circuits, the opamp must be reset to discharge the next stage sampling capacitor, adding the opamp noise in phase 1 to the next stage. This noise will be shaped mainly by the bandwidth of the next stage in the amplification phase and thus will be calculated as part of the MDAC noise in next Section.

During phase 2 (Fig. 5.7b), both the opamp input referred noise density and the on-resistance thermal noise density of the switches will be filtered by the closed-loop bandwidth in phase 2. The noise from the opamp in phase 2 can be calculated as:

$$\overline{v_{n,opamp}^2[\phi2]} = S_{opamp}(0) \cdot |H_{\phi2}(0)|^2 \cdot \frac{1}{4}\beta\omega_{GBW\_\phi2}$$

$$= \frac{1}{4\beta}S_{opamp}(0)\omega_{GBW\_\phi2} \qquad \text{(C4)}$$

where $H_{\phi2}(0) = 1/\beta$ is the low-frequency noise voltage transfer function from the opamp input referred noise source to the output in phase 2, β is the feedback factor in phase 2, $\omega_{GBW\_\phi2}$ is the gain-bandwidth product of the opamp. The noise from the switches in phase 2 can be evaluated similarly:

$$\overline{v_{n,Rs,B}^2[\phi2]} = 4kT(R_{s1,B} + R_{s2,B}) \cdot |H_{C2,\phi2}(0)|^2 \cdot \frac{1}{4}\beta\omega_{GBW\_\phi2}$$

$$= kT(R_{s1,B} + R_{s2,B})\beta\omega_{GBW\_\phi2} \qquad \text{(C5)}$$

Thus the total noise power in fully-differential mode is twice the sum of all the above contributions (with $\beta = 1/3$):

$$\overline{v_{n,S/H}^2} = 2\left\{ 2kT/C + \omega_{GBW\_\phi1}\left[kTR_{s2,A} + \frac{1}{2}S_{opamp}(0)\right] \right.$$

$$\left. +\omega_{GBW\_\phi2}\left[\frac{1}{4\beta}S_{opamp}(0) + kT\beta(R_{s1,B} + R_{s2,B})\right]\right\} \qquad \text{(C6)}$$

MDAC with Auxiliary Amplifier

This section provides a thermal noise analysis of the MDAC circuit with auxiliary amplifier used for finite-gain compensation, as shown in Fig. 5.4. In phase 1 (with equivalent circuit in Fig. 5.8a), the thermal noise sampled in the capacitors C_{ref} and C_{f1} will be transferred to the output in phase 2:

$$\overline{v_{n,kT/C}^2[\phi1]} = kT/C_{ref}\left(\frac{C_{ref}}{C_{f1}}\right)^2 + kT/C_{f1} \qquad \text{(C7)}$$

again the noise sampled in capacitor C_{f1} is still equal to kT/C_{f1}.

Special attention must be devoted to the handling of the noise from the main opamp, since this noise will be fed into the auxiliary opamp. In phase 1, the main opamp output node is connected to the bottom plate of capacitor C_{s2}, but, simultaneously, this noise is fed to the virtual ground of the auxiliary amplifier and then sampled into the top plate of C_{s2}, thus partially cancelling the main opamp noise contribution to C_{s2}. The reason for this partial cancellation is due to the extra auxiliary opamp bandwidth. To evaluate the noise of the main opamp, sampled in capacitor C_{s2}, it would be necessary to evaluate the noise transfer function from the main opamp virtual ground to the differential voltage of C_{s2}:

$$\frac{V_{G2} - V_{o1}}{V_1}(s) = \frac{1}{\left(1 + s/\omega_{GBW\_main\_\phi1}\right)\left(1 + s/\omega_{GBW\_aux\_\phi1}\right)} - \frac{1}{1 + s/\omega_{GBW\_main\_\phi1}}$$

$$= \frac{-s/\omega_{GBW\_aux\_\phi1}}{\left(1 + s/\omega_{GBW\_main\_\phi1}\right)\left(1 + s/\omega_{GBW\_aux\_\phi1}\right)}$$

$$(C8)$$

Then, the noise power sampled in capacitor C_{s2} is equal to

$$\overline{v_{n,(V_{G2}-V_{o1})\_main}^2}[\phi1] = \frac{1}{2\pi}\int_0^\infty S_{main}(f)\left|\frac{V_{G2} - V_{o1}}{V_1}(j\omega)\right|^2 d\omega$$

$$= \frac{1}{4}\frac{S_{main}(0)\omega_{GBW\_main\_\phi1}^2}{\omega_{GBW\_main\_\phi1} + \omega_{GBW\_aux\_\phi1}}$$

$$(C9)$$

On the other hand, the noise power sampled in capacitor C_{f2} can be evaluated similarly:

$$\overline{v_{n,V_{G2}\_main}^2}[\phi1] = \frac{1}{2\pi}\int_0^\infty S_{main}(f)\left|\frac{V_{G2}}{V_1}(j\omega)\right|^2 d\omega$$

$$= \frac{1}{4}\frac{S_{main}(0)\omega_{GBW\_main\_\phi1}\omega_{GBW\_aux\_\phi1}}{\omega_{GBW\_main\_\phi1} + \omega_{GBW\_aux\_\phi1}}$$

$$(C10)$$

The noise power (from the main opamp in phase 1) is then transferred to the output in phase 2 as:

$$\overline{v_{n,main\_opamp}^2}[\phi1] = \frac{1}{4}\frac{S_{main}(0)}{\omega_{GBW\_main\_\phi1} + \omega_{GBW\_aux\_\phi1}}$$

$$\times \left\{\omega_{GBW\_main\_\phi1}\left(\frac{C_{s2}}{C_{f2}}\right) + \sqrt{\omega_{GBW\_main\_\phi1}\omega_{GBW\_aux\_\phi1}}\right\}^2$$

$$(C11)$$

Also, the noise of the auxiliary amplifier in phase 1 is sampled in both C_{s2} and C_{f2} and transferred to the auxiliary output in phase 2 as:

$$\overline{v^2_{n,aux\_opamp}}[\phi 1] = S_{aux}(0)\frac{1}{4}\omega_{GBW\_aux\_\phi 1}\left[1+\frac{C_{s2}}{C_{f2}}\right]^2$$

The main opamp resets in phase 1 and generates noise to the next stage, however this noise is shaped by the more dominant next stage bandwidth and thus it will be counted as the noise in next stage.

During phase 2 (Fig. 5.8b), the noise from the three resistors can be evaluated as:

$$\overline{v^2_{n,Rref,B}}[\phi 2] = 4kTR_{ref,B}\cdot\left|H_{ref,B,\phi 2}(0)\right|^2\cdot\frac{1}{4}\beta\omega_{GBW\_main\_\phi 2}$$

$$= kTR_{ref,B}\left(\frac{C_{ref}}{C_{f1}}\right)^2\beta\omega_{GBW\_main\_\phi 2} \tag{C12}$$

$$\overline{v^2_{n,Rs2,B}}[\phi 2] = 4kTR_{s2,B}\cdot\left|H_{Cs2,\phi 2}(0)\right|^2\cdot\frac{1}{4}\beta\omega_{GBW\_aux\_\phi 2}$$

$$= kTR_{s2,B}\left(\frac{C_{s2}}{C_{f2}}\right)^2\beta\omega_{GBW\_aux\_\phi 2} \tag{C13}$$

$$\overline{v^2_{n,Rf2,B}}[\phi 2] = 4kTR_{f2,B}\cdot\left|H_{Cf2,\phi 2}(0)\right|^2\cdot\frac{1}{4}\beta\omega_{GBW\_aux\_\phi 2}$$

$$= kTR_{f2,B}\beta\omega_{GBW\_aux\_\phi 2} \tag{C14}$$

Again, the noise from the main opamp should be treated carefully since the noise is partially cancelled in C_L:

$$\frac{V_{o1}-V_{o2}}{V_1}(s) = \frac{1/\beta}{1+s/\beta\omega_{GBW\_main\_\phi 1}} - \frac{1/\beta}{\left(1+s/\beta\omega_{GBW\_main\_\phi 1}\right)\left(1+s/\beta\omega_{GBW\_aux\_\phi 1}\right)} \tag{C15}$$

$$\overline{v^2_{n,(Vo1-Vo2)\_main}}[\phi 2] = \frac{1}{2\pi}\int_0^\infty S_{main}(f)\left|\frac{V_{o1}-V_{o2}}{V_1}(j\omega)\right|^2 d\omega$$

$$= \frac{1}{4\beta}\frac{S_{main}(0)\omega^2_{GBW\_main\_\phi 2}}{\omega_{GBW\_main\_\phi 2}+\omega_{GBW\_aux\_\phi 2}} \tag{C16}$$

The noise from the auxiliary opamp is

$$\overline{v^2_{n,aux\_opamp}}[\phi 2] = S_{aux}(0)\frac{1}{4}\left(\frac{1}{\beta}\right)^2\beta\omega_{GBW\_aux\_\phi 2}$$

$$= S_{aux}(0)\frac{1}{4\beta}\omega_{GBW\_aux\_\phi 2} \tag{C17}$$

The noise from the previous reset stage (with noise PSDs $S_{pre}(f)$) is also shaped by the bandwidth of this stage:

$$\overline{v_{n,pre}^2[\phi 2]} = S_{pre}(0)\left|H_{Cs1,\phi 2}(0)\right|^2 \frac{1}{4}\beta\omega_{GBW1\_\phi 2}$$

$$= \frac{1}{4}S_{pre}(0)\left(\frac{C_{s1}}{C_{f1}}\right)^2 \beta\omega_{GBW\_main\_\phi 2} \tag{C18}$$

Thus, the total noise of the MDAC (in fully-differential mode) can be calculated as:

$$\overline{v_{n,MDAC}^2[\phi 2]} = 2kT/C_{ref}\left(\frac{C_{ref}}{C_{f1}}\right)^2 + 2kT/C_{f1}$$

$$+ \frac{S_{main}(0)}{2}\left\{\frac{\left[\omega_{GBW\_main\_\phi 1}\left(\frac{C_{s2}}{C_{f2}}\right) + \sqrt{\omega_{GBW\_main\_\phi 1}\omega_{GBW\_aux\_\phi 1}}\right]^2}{\omega_{GBW\_main\_\phi 1} + \omega_{GBW\_aux\_\phi 1}}\right.$$

$$\left. + \frac{\omega_{GBW\_main\_\phi 2}^2/\beta}{\omega_{GBW\_main\_\phi 2} + \omega_{GBW\_aux\_\phi 2}}\right.$$

$$+ \frac{1}{2}S_{aux}(0)\left\{\omega_{GBW\_aux\_\phi 1}\left[1 + \frac{C_{s2}}{C_{f2}}\right]^2 + \frac{1}{\beta}\omega_{GBW\_aux\_\phi 2}\right\}$$

$$+ \frac{1}{2}S_{pre}(0)\left(\frac{C_{s1}}{C_{f1}}\right)^2 \beta\omega_{GBW\_main\_\phi 2}$$

$$+ 2kT\beta\left\{R_{ref,B}\left(\frac{C_{ref}}{C_{f1}}\right)^2 \omega_{GBW\_main\_\phi 2}\right.$$

$$+ \left[R_{s2,B}\left(\frac{C_{s2}}{C_{f2}}\right)^2 + R_{f2,B}\right]\omega_{GBW\_aux\_\phi 2}\right\} \tag{C19}$$

Appendix D
Special Case in Gain Mismatch

This appendix provides the proof to the observed phenomena measured in the six-channel un-calibrated output spectrum due to a special case of gain-mismatch highlighted in Chapter 6. In its general form the output spectrum of the time-interleaved ADC with a sinusoidal input signal under gain-mismatch consists of modulation sidebands with weights as presented in (4.23) (with $H(\omega) = 1$):

$$A_k = \frac{1}{M} \sum_{m=0}^{M-1} (1 + \delta_m) e^{-jkm\frac{2\pi}{M}} = \frac{1}{M} \sum_{m=0}^{M-1} a_m e^{-jkm\frac{2\pi}{M}} \tag{D1}$$

where $a_{\mathrm{m}} = 1 + \delta_{\mathrm{m}}$. Under special layout conditions (illustrated in Fig. 5.9) it can be assumed that

$$a_m = a_{M/2+m} \tag{D2}$$

since the gain mismatch is minimized between the channels in close proximity (for those sharing the same clock buses). The weights A_k can then be expressed as:

$$A_k = \frac{1}{M} \sum_{m=0}^{M/2-1} a_m \left[e^{-jkm\frac{2\pi}{M}} + e^{-jk(\frac{M}{2}+m)\frac{2\pi}{M}} \right]$$

$$= \frac{2}{M} \cos\left(\frac{k\pi}{2}\right) e^{jk\pi/2} \sum_{m=0}^{M/2-1} a_m e^{jk\pi/2} e^{-jkm\frac{2\pi}{M}} \tag{D3}$$

which are non-zero only for even values of k. If $M = 6$ then $A_k \neq 0$ for only $k = 0$ (the signal components) and $k = 2$ (the 120 MHz $\pm f_{in}$ modulation sidebands as shown in Fig. 6.7), which confirms the results reported in Fig. 6.7. Also, the component at $k = 4$ is the image component of that of $k = 2$.

Index